Understanding
Engineering Design

Understanding Engineering Design

Context, Theory and Practice

Richard Birmingham
Graham Cleland
Robert Driver
David Maffin

PRENTICE HALL
London   New York   Toronto   Sydney   Tokyo   Singapore
Madrid   Mexico City   Munich   Paris

First published 1997 by
Prentice Hall Europe
Campus 400, Maylands Avenue
Hemel Hempstead
Hertfordshire HP2 7EZ
A division of
Simon & Schuster International Group

Typeset in 11/13pt Plantin
by Hands Fotoset, Leicester

Printed and bound in Great Britain by
T.J. Press (Padstow) Ltd

Library of Congress Cataloging-in-Publication Data

Understanding engineering design : context, theory, and practice /
  Richard Birmingham . . . [et al.].
     p.   cm.
  Includes bibliographical references and index.
  ISBN 0-13-525650-X (hbk. : alk. paper)
  1. Engineering design.   I. Birmingham, Richard.
TA174.U48   1996
620′.0042—dc20   96-19064
                                                           CIP

British Library Cataloguing in Publication Data

A catalogue record for this book is available from
the British Library

ISBN 0-13-525650-X

---

1 2 3 4 5   01 00 99 98 97

Contents

Plates

Figures

Figures

Foreword by
Professor Peter Hills

This book is timely: modern industrial countries must understand and exploit modern engineering design. They must understand its *context*, its *theory* and its *practice* if they are to realise its full potential. If a country has the intellectual resources to design what its people want and the will to invest in the creation of those products, then it can avoid exploitation by others and provide the wealth that can ensure a happy, healthy and well-educated population. It raises itself above being merely the provider of cheap labour and gives its people the means to reap the prime benefits of modern industry.

However, world-class performance in design and new product introduction means that all the present possibilities and future potentials of engineering design must be understood. This can only happen if attention is paid to the thoughts of people like these authors, whose researches have helped us to our present design capabilities.

When I started work in a design office in the 1950s I was one of some two hundred and fifty design draughtsmen. We used pencils, drawing boards and slide rules. Our methods were established practice and company standards. Later, I worked on early computer-aided design, using a computer which, with its requisite air conditioning and all the staff needed to coax it into obliging us, filled a converted fire station.

Now, the factory where I started employs about forty engineers using computers to produce more designs in less time than we did; and they are more confident about the quality and performance of what they are designing than we were. They can visualise, analyse, optimise, and simulate, and are close to the point where they can create a virtual product in the computer, test it against the needs and wishes of all who will come into contact with the product, ensure that all legal requirements are met, and then translate it into affordable and reliable reality, without ever prototyping it.

When any of those engineers goes on a business trip, he or she will almost certainly be carrying a lap-top computer, the memory and speed of which so far outstrip the capabilities of that earlier machine that to compare them is like comparing a modern materials-handling robot with a crowbar.

In forty years, designing artefacts to compete internationally has changed beyond recognition. The management of design has changed, making lead times a fraction of what they once were, while the ability to design for a whole range of functional criteria, for optimal manufacture and ease of assembly, yields engineering products and consumer durables which are cheaper but better than ever before.

The authors have put their thoughts in print in a useful and practical way. This book has much to say to managers, engineers, and designers, whether at the bottom or the top of their professional ladders. It would benefit the interested lay person by providing an insight into a fascinating world of responsible creativity. It does not shirk saying what is difficult. It describes modern methods and puts them in a meaningful industrial and historical context. It presents its message conveniently and clearly, with summaries at the end of each chapter to help the busy reader home in quickly on areas of immediate concern. Above all, I believe, it will pay handsome dividends to the reader.

Professor Peter Hills
President of the Institution of Engineering Designers
January, 1996

Preface

Engineering design became a recognised subject of academic study in the early 1960s. Since that time a great deal of work has been carried out by researchers of many nationalities, and many papers have been published in journals and presented at conferences. In addition several influential texts have been written, a few of which are philosophical in nature, while most describe their author's view of the design process, both in theory and in practice. As these books range in style from the weighty treatise to the readily accessible manual, it must be asked, what can this text bring to the field?

The four authors of this text met while working on separate design-related research projects at the University of Newcastle upon Tyne. Each had a background which included a degree in an engineering subject, together with industrial and design experience. Yet despite this all felt ill prepared for their research tasks, and considered further study in the field of engineering design essential. This common desire to expand their understanding of the subject led to the creation of an informal study group which, over a twelve-month period, read and discussed the key texts and papers in the field. From this comprehensive review of the literature it emerged that while each text provided a sound basis for the practice of engineering design, none presented a rounded overview of the field which encompassed its diversity and complexity of internal debate. This is the contribution that it is hoped this text can make.

To this subject the four authors bring a wide diversity of experience. David Maffin is a senior researcher in the Centre for Urban and Regional Development Studies at Newcastle University. He has a general interest in technological change and innovation in manufacturing and, in particular, he has undertaken research into engineering design and product development. He previously spent eleven years in industry, working in the area of design and project management. Robert Driver studied engineering at Cambridge University, since when he has been a development engineer with Parsons Power Generation Systems Ltd. (part of the Rolls-Royce Industrial Power Group). His recent research has been into methods of modelling interconnected systems. Graham Cleland is a design and projects manager with Procter and Gamble, having obtained his doctorate studying the application of artificial intelligence systems to engineering design, this work being sponsored by BP, Shell and British Gas. Richard Birmingham is a lecturer in Small Craft Design in the Department of Marine Technology at Newcastle University. His doctoral studies were into the concept stage of structural design, and followed ten years of work as a boat-builder and designer. All four have contributed papers in the field of design.

Combining this wide variety of backgrounds in both engineering and design, the authors set contemporary design theory in the context of engineering organisations and design practice. While the book can stand alone as an introduction to the subject, the intention is that it should be a primer for further reading. References throughout the text, together with the comprehensive bibliography, provide a guide to sources where more detailed information on all aspects of engineering design can be found. This text should therefore provide a useful foundation for undergraduates in all engineering disciplines, and an insight into the field for postgraduates embarking on design related research. For practitioners of design the broad overview and detailed bibliography should prove a useful resource.

Introduction

design process
Manufacturing process

Good product design is fast becoming recognised as a basis for securing competitive advantage for manufacturing business. In recent years the subject of design and product development has been the focus of government initiatives in the UK and elsewhere which have been aimed specifically at bringing about business performance improvement. Many individual organisations have themselves begun to realise the benefits that effective and efficient management of the design process can bring to a business. For many years, organisations tended to concentrate their efforts on improving the functional performance of their respective manufacturing processes. This focus yielded many gains, but nowadays the commonly held belief is that radical improvements can be realised only through better organisation, management and control of the design process, as well as the manufacturing processes. It is evidently the case since much design activity, particularly of the more detailed type, is actually a form of tactical planning for manufacturing. Of course, it could equally be argued that the more esoteric types of design activity are, at least to a certain extent, planning for manufacturing too, albeit that this is of a more abstract and conceptual character. Nonetheless, since there is currently a strong degree of unanimity about the importance of design, it is necessary to consider briefly why this should be the case.

Over the years, since the days of the industrial revolution in the UK, much has been written about design, and while some of these writings are still useful some are not. The literature relating to the subject dates back a long way, but it is only in the last thirty or forty years that the subject has become widely recognised as a discipline in its own right. This body of literature covers many diverse aspects of design and some of the more popular topics that have been the frequent subject of debate have included problem solving, design process modelling and professional ethics. Many of the more important contributions are summarised in this text, each chapter including references to publications mentioned in the discussion.

1

Full bibliographic details of all these references can be found in the extensive bibliography, which also includes other noteworthy contributions to the field which could not be dealt with in this volume.

The topics contained in this text were selected for inclusion on the basis that each represents a different perspective of design. Because the text is intended as a primer, the presentational format adopted is unlike that of many of the other more prescriptive design texts available. The intention is not to tell the reader how to do design, but to inform the reader as to the nature of design. As such, the text takes a holistic and pragmatic view of the role that a designer plays in a design project.

The book itself is divided into three sections, each dealing with one of the principal areas of interest to designers. The first section of the text deals with the context within which design is, or possibly may be, carried out. Nowadays manufacturing is a very competitive business environment which demands that organisations must be lean, agile, innovative and dynamic in order to survive. Close scrutiny of the nature of manufacturing sector markets, and hypothesis of the potential future trends in these markets, seems to support the notion that the over-capacity, under-utilised and slow-response businesses will be unable to react quickly enough to changing market conditions, and are therefore unable to satisfy their market's needs within a competitive period of time. Design, in particular, is an activity that cannot be carried out in a vacuum since there are many important factors that always need to be taken into account in evolving a successful solution to a problem. Accordingly, it is important to have an understanding of the environmental influences which come to bear on a design project, but over which a designer has no control. These environmental influences might be due to technological advances, changes in the socio-economic climate, or changes in the physical setting within which a designer operates. In fact, a hierarchy of influences exists and all designers must develop their knowledge of this hierarchy and how the individual environmental influences might impact upon their work.

In addition to understanding the impact that certain influences might have on a design process, designers must also develop an understanding of the role that they play in the process. As well as performing the various activities necessary to enable the transition from identification of a problem to generation of a successful solution, designers should be aware of the need to act as a communicator, project manager, professional and responsible engineer, and so forth. Of course, one of the most crucial roles of the designer is that of innovator. Manufacturing business is becoming dominated by the need continually to introduce new and innovative products, even to markets which traditionally have stable products. Designers must understand that an innovation need not be applied to a product itself, but might be in the provision of after sales support, or in the recycleability of a product. As the need for businesses to be more innovative has gradually increased over time, so has the need for designers to possess the ability to identify and then subsequently manage the associated risks. Until recently, risk-related issues would have been very low on the typical designer's agenda, but there is an increasing need both to comprehend the nature of risks associated with innovation, and to understand how these risks can be minimised.

The second section of the text is focused on the history of developments in design that have occurred over the last two hundred years or so. The entire nature of engineering business has shifted during this period, companies moving from relying on a single expert, called the craftsman, who possessed all the knowledge necessary to develop a successful solution to a given problem, to the concept of a multidisciplined team who can simultaneously apply their collective know-how to solving design problems. The developments discussed also include those related to recent advances in design theory, and many of the relevant contributions are discussed at length. Over this period the type of equipment that designers have at their disposal has also changed markedly. Computers play an important and influential role in life today, and in design are becoming virtually indispensable. They can be employed to assist in the entire process of taking a problem through from identification of a need, to development of a satisfactory solution. Not only are such devices able to aid in the process of swiftly generating alternative proposals, evaluating these and selecting a favoured candidate, and then developing this as an embodiment of an actual product, but they are equally able to assist in ensuring that the product can be operated safely, maintained easily and disposed of readily. What is more, computers today can also play a significant part in manufacturing the physical artefacts themselves, thereby further encroaching on the role of the traditional craftsman.

The final section of the text is specifically focused on the practice of design. Newcomers to the design field have to develop their understanding of the sorts of tools, techniques and methods that are available which can assist in progressing the design process. Furthermore, it is important to develop a complete and comprehensive understanding of the strategies that can be adopted to help in the process of solving particular problems.

3

Indeed, it has been said that the first step a designer must take in a design project is to design the design process itself. Appropriate selection of the approach that will be adopted for a particular project can go a long way to ensuring the success of that project. Consequently, designers must develop their knowledge of the options open to them and the viability of employing these different options in different circumstances. This section also considers the raw material of design, which is information, as in the many diverse roles that designers play the common thread is the generation, manipulation, transformation, and communication of information. Designers must learn how to manage properly this information through the life of a project because it is the effective and efficient management of this information which actually ensures the viability of a business. Successfully managing and controlling the information explosion, which inevitably occurs during the design process, is a key role of the designer.

The subjects covered in this text include those that have been discussed extensively during the last thirty or forty years. Inevitably the discussion is not exhaustive, but it is hoped that it will provide an insight into the challenges of taking up design as a career, and communicate both the enjoyment that can be derived from delivering satisfaction to others and the fulfilment that can be gained from working at the forefront of technological advances.

Design is not easily defined in a concise, easily understood format. Indeed, defining design, or simply providing a concise description of what it involves, is a very subjective exercise. As such, there will always be arguments as to which topics could be included in addition, or in preference, to those covered in this text. As an introductory guide to design, however, the authors believe that the text serves its purpose well, and indicates the diverse scope of opportunities available within a design career. While some practising designers derive their satisfaction from seeing their own ideas become real objects, possibly used in everyday life by people the world over, others are content to be working at the forefront of technology, fully aware that they may seldom see their creative efforts get beyond the drawing board. In either case, the activity of design, with its need for creativity on the one hand, and attention to detail on the other, is extremely rewarding.

Section I
The Context of Engineering Design

Chapter 1
The environment of design

Introduction

A designer's surroundings can be described in a variety of ways. The factors which can influence a designer's thoughts, views and actions can be physical in nature, but they can also be more abstract, such as political or philosophical. In this opening chapter four areas which can have an influence on design activity are described. First, the pace of technological development is considered, as it is this which has helped place design in its position of importance today. Second, the structure which defines a designer's responsibilities is examined, before describing the physical setting in which those responsibilities are shouldered and work is carried out. From within that setting the designer has to interact with colleagues, customers, and the rest of society, so the fourth view is the economic and social context within which the designer operates. Together these different perspectives make up the environment of design.

The technological environment and the accelerating rate of change

Stone Age people were faced with many uncertainties, but in some aspects their life was remarkably stable. Tools recovered from prehistoric times show little change in style over many thousands of years. Evidently a flint worker would produce the same pattern of arrowhead as his father, and this satisfactory pattern would be passed down over many generations. Even in the comparatively recent Middle Ages, items such as agricultural tools remained the same for centuries. This technological stability contrasts markedly with the technological turbulence which is a characteristic of modern times. When conditions change, individuals and society find that they have new needs. The changing requirements encourage new developments, so innovation is rewarded and new artefacts are created.

One way of perceiving the activity of design, therefore, is as a response to the emergence of new needs. But for the response to be appropriate the precise nature of the new need must be identified. This can be demonstrated with the following example. If, over a period of years, an island becomes increasingly populated, it may eventually become evident that a bridge linking it with the mainland is required. The need is obvious: a bridge. But what sort of bridge should be built? Should it be a foot bridge or a road bridge, and in either case what will be the density of traffic? Furthermore, do boats have to pass under the bridge, and if so are they small motor boats, or tall yachts, or even ships? Such questions clearly indicate that the precise nature of the need must be analysed before any solution can be found. Converting perceived needs, which may be vaguely described, into concrete design *goals* or *objectives* focuses the design activity into satisfying set *requirements*. It is this process of *problem definition* which ensures that the design effort is properly directed, and that its success can be demonstrated.

If the design process is to be successful, then problem definition must be rigorous. Indeed this is such an important activity that an entire profession has evolved, dedicated to identifying individual or group needs. This activity is called marketing, and it is worth noting that marketing can be used either to identify a need (and so direct the design activity), or to create artificially a need for designed products which would otherwise be considered unrequired.

The relationship between needs and new designs is actually more subtle than that of cause and effect. Design activity does not only satisfy requirements, but is itself the source of new ones. This is a cycle in which change can lead to new needs, needs lead to new designs, and these lead to more change, as shown in Figure 1.1. Since the Stone Age this cycle

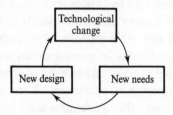

Figure 1.1   The accelerating cycle of technological change.

of technological change has been accelerating, to the extent that today some Japanese manufacturers of electronic equipment complete the product life cycle of invention, manufacture, and obsolescence in six months. Compared to the technological stability of the past our times are indeed turbulent, but it is this which has made design so important in the modern age.

The organisational environment and the hierarchy of design decisions

Decision making and design are so intertwined that it has been suggested that all decision making can be viewed as design (Simon, 1969). Certainly design activity can be characterised as a series of decisions. These gradually refine a concept, which may at the outset be almost abstract, by defining different aspects of its character until a specification is created which is detailed enough to allow manufacture. These defining decisions can be seen as progressively introducing limits or *constraints* on the possible solution, until it is devoid of ambiguity.

This view of the design process can be explored further if we return to the example of a bridge linking an island to the mainland. We have already indicated that it must be decided if it is to be a foot or road bridge, and what vessels must be able to pass under it. If it is decided that it must be a bridge with enough clearance for a yacht's masts then we have set constraints which immediately exclude many other possible solutions, such as a causeway penetrated by pipes which allow only the passage of water. The next set of decisions might be to establish what form the bridge should take, and which material should be used. The possibilities may include a stone arch, a concrete or steel beam, a steel truss, or even a suspension bridge, as shown in Figure 1.2. Having decided on these we may progress to deciding what should be the overall configuration, perhaps how many arches, or the layout of the truss, and then move on to the detailed dimensions of individual components, and how these are to be joined. It can be seen that each decision defines the solution more clearly, but that it also introduces constraints, and so reduces the alternatives from which the final solution is to be found. For example, if it is decided that it is to be a suspension bridge, then the material must be able to support tensile loads, so the bridge cannot be made of stone or concrete since these materials have poor tensile properties.

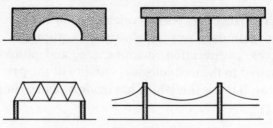

Figure 1.2   Alternative bridge types.

The simplified description given above is in one way misleading, for the process is often not a linear progression through a list of decisions. Before any one decision can be finalised it may be necessary to consider the ramifications of that particular choice. For example we may decide on a stone arch, only to discover that it is impossible to produce a span of adequate size. Tentative proposals therefore have to be put forward, the implications clarified, and then the initial decision reconsidered. Often designers have to repeat the same steps many times, which is why designing is always described as highly *iterative* in nature.

The design decisions can be arranged as a hierarchy. The decisions made at one level impose constraints on the next level down, so they effectively set the framework within which the next set of decisions are allowed to be made. In some cases different people make the decisions at each level of the hierarchy. For example the senior designers may decide the bridge is to be a truss, and give that as the framework for more junior designers to establish the most satisfactory layout of components. This layout may in turn be passed on to trainee engineers to calculate the exact dimensions of each of these individual components. At each level of the hierarchy the decisions are passed down as a framework within which the next level must work.

The hierarchy of design decisions can actually be extended beyond the consideration of technical aspects of the problem. Economic and financial implications will clearly influence the final design, and in many cases social or political pressures are also important. These elements influence the design process at the highest levels of the decision hierarchy. The example followed above starts with the assumption that it is a bridge that is needed, but this assumption could itself be a constraint imposed from above. Consideration could be given to alternative ways of linking the island, such as a ferry or tunnel. And this decision still need not be the peak of the hierarchy, for an even wider question could be asked: should the islanders be rehoused on the mainland instead of creating a link at all?

Clearly, many of these decisions lie outside the designer's realm, but the designer must be aware of this hierarchy, for in one sense this sets the environment in which the designer works. At each level of the decision hierarchy the scope for decision making is defined, and different types of consideration influence the outcome. At the highest levels the pressures can be political, social, economic, financial, or competitive (Wallace and Hales, 1987). Although these can properly be called design decisions, they are not necessarily made by designers but by others such as managers or politicians. These decision makers may still choose to involve designers, but in the role of consultants or advisors. This view of the decision hierarchy can be represented as a series of concentric rings, each of which defines the boundaries and responsibilities for those working within it, as shown in Figure 1.3. As designers progress in seniority, so their scope for decision making increases, but for every designer or decision maker, the contextual environment is defined by those above.

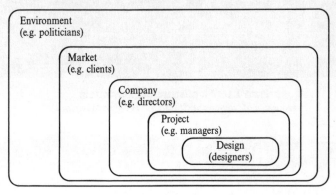

Figure 1.3   Examples of decision makers in the design decision hierarchy.

The physical environment and the designer's setting

The products that an engineering designer could be associated with encompass a wide range of artefacts (Plates 1.1 to 1.4), some of which require design activity on a far larger scale than others. At one end of the spectrum are huge one-off products such as hydroelectric dams, power stations, or offshore production platforms. These require a massive and skilfully organised design effort. Other industries operate their design activity on a similar human and financial scale, even if the end-product is smaller and not unique. Two contrasting examples of this case are jumbo

Plate 1.1　An offshore exploration rig
(courtesy of Newcastle Chronicle and Journal Limited).

Plate 1.2　A passenger jet aeroplane
(courtesy of Newcastle Chronicle and Journal Limited).

Plate 1.3   A computer

Plate 1.4   A mountain bike
(courtesy of Newcastle Chronicle and Journal Limited).

jets and microprocessors. Many designers, however, are involved with objects for which the design activity is on a more modest (yet still significant) scale, such as cars, computers and televisions. Finally some products, such as a bicycle or children's toy, can be designed by a single person working unaided.

These contrasting design situations result in an equally wide range of design settings within which the designer may operate (Plates 1.5 and 1.6). In some cases the designer could be working from home, squeezing into the spare bedroom a drawing board, personal computer, and fax machine. Alternatively, in a small company, the designer may carry a variety of responsibilities, including both the design of the product and specifying the process which will be used to produce it. In larger companies design activity is frequently not carried out by a single person, but requires a design team. The staff in the design office will each have specified areas of activity and levels of responsibility, and will be generally supported in administrative activities by secretarial and clerical staff. Finally, very large organisations may concentrate considerable numbers of designers into one building, or alternatively may spread the design function over many sites. In this case in addition to administrative help, they may have large drawing offices, powerful CAD tools (software packages for *computer-aided design*) and efficient communications systems to all parts of the world.

It should be realised that the link between the individual designer and the organisation requiring the design may take a variety of forms. Clearly the organisation may directly employ the designer as part of its own design team but, alternatively, all or part of the design may be commissioned from an outside company of design consultants. In some cases an organisation may augment its own team by using designers seconded from an external firm of consultants, in which case the individual is employed by one company, but working for another. The nature of this link can affect the design activity, for external consultants may have less loyalty to existing designs and processes which have been used by the company in the past. They have a stronger tendency to look to wider horizons for inspiration and new design solutions. In contrast, the designer who is directly employed by the organisation may well have invested considerable effort into the existing designs, and will certainly have access to historical data concerning their performance. This will inevitably encourage a tendency to stay with solutions to design problems that have been 'tried and tested'.

Regardless of whether the designer is working alone or as part of a

Plate 1.5   A designer at work
(courtesy of Newcastle Chronicle and Journal Limited).

Plate 1.6   A large design office
(courtesy of Parsons Power Generation Systems Limited).

team, or whether the organisation employs the designer directly or tenuously, it is the case that much of the equipment in the design office will be concerned with the manipulation of information. Information is the basic material of the design activity, and the designer has to gather it from available sources, process it into a useful form, and finally deliver it neatly packaged to a customer. This perspective will be discussed further in the third section of this book. At this stage it is simply important to appreciate that the designer is a manipulator of information, and that the designer's tools are therefore primarily concerned with this activity. Some are for communicating information, such as the telephone, word processor, fax machine, and computer links. These are used to gather information and to pass it on to other members of the design team, or to the customer. Other tools such as computer software packages, or even pocket calculators, are concerned with the manipulation and evaluation of information, and with generating useful data from the raw material. Lastly drawings, whether the product of the traditional drawing board or of today's sophisticated graphics packages, are not only concerned with communicating the final design, but also with storing information in a readily accessible form. As with other craftsmen, designers are surrounded by the tools of their trade, but today these include the telephone, computer, fax machine, and so on.

The socio-economic environment and the justification of design

In the discussion above it was shown how design activity is instigated in response to perceived needs, but it must be recognised that not every need leads to a new design. The design process can be a costly one as it inevitably takes time. People are an expensive resource, and so committing their time to satisfying a specific requirement must be justified. Not all needs can be responded to, and therefore mechanisms are required to select those which should be addressed and those which should be ignored.

In our society one mechanism which effectively focuses the design effort into specific projects is the profit motive. The commercial drive for a business to remain solvent, and preferably to prosper, dictates that companies carefully examine alternative design options before embarking on a major project. Typically they will quantify the need in financial terms, and assess the probability that they will receive in earnings a large enough proportion of that total to more than pay for their own investment. The precise form of this assessment exercise, which is part of the marketing process, varies according to the nature of both the product and the customer. The needs of an entire sector of society will be examined using questionnaires and market surveys before embarking on designing an item for mass production, while key individuals in selected professions or companies may be interviewed in preparation for a high cost/low volume product. The most extreme case may be found with large made-to-order products, such as a power station or chemical plant, where identifying the precise 'need' and its value may be negotiated by a team of sales executives over many months.

Even if these exercises establish a substantial market for a proposed product, other considerations may act as disincentives to embarking on a project. Risks which may be of many different types must be assessed. These risks include unexpected technical difficulties both in the functioning of the product and in its manufacture. There is also the possibility of changes in the commercial situation due to outside influences or competitors' actions. This is particularly problematic in projects for which the time lapse between the identification of a need and the realisation of a practical solution is long, as for example in the case of building the Channel Tunnel, discussed in the next chapter. Lastly, political developments such as unfavourable legislation or civil unrest could jeopardise the project. Clearly a trade-off is necessary between the magnitude of the risk and the size of the anticipated return. If the risk and reward equation is perceived as unacceptable, design activity will be suppressed.

The profit motive is not the only mechanism which can trigger design activity. When there is no commercial justification to embark on a design exercise, external agencies can deliberately intervene to ensure design activity occurs and is aimed in a particular direction. One way that this can be achieved is by simply declaring that there is a particular requirement, and announcing that funds will be made available to finance the necessary design response. In many instances it is governments which commission projects in this way, to provide the national infrastructure of roads, railways, sewage systems, and other services. Governments also instigate design for defence equipment and for prestige projects which can bring international recognition, such as a national monument, or Concorde, the supersonic airliner. Non-governmental organisations can also be the agents for this type of activity, acting either directly or in more

17

subtle ways. The major charities, for example, act to instigate design which will assist in their specific mission, be it to provide shelter, water, or other services to deprived communities. Pressure groups such as Greenpeace or Friends of the Earth act indirectly, aiming to alter public opinion, or heighten public awareness, and so create a market demand for newly perceived needs. Lead-free petrol is an example of a new requirement created in this way. The design effort which eventually produced a satisfactory combination of lead-free petrol and appropriately tuned engines was the result of a combination of the mechanisms mentioned above. Pressure groups altered public opinion, forcing government intervention in the form of tax incentives. These ensured that the profit motive was activated and produced an appropriate design response.

From the above discussion it is clear that the designer is subjected to many influences of which some are direct, while others are more subtle. Some have a moral dimension and encourage designers to consider the wider implications of their activity by looking beyond the company cash flow to consider the true cost and benefits to society. In many cases the difficult environmental and social implications cannot be ignored. Laws, regulations, and codes of practice force these issues to be acknowledged, while grants and subsidies may encourage more socially acceptable courses of action.

Summary

The physical environment in which the engineering designer works can range from a cramped office, with little or no administrative support, to extensive offices housing large design teams who are well provided with clerical and administrative assistance. Despite this diversity in their surroundings, the working environments of all designers can be shown to have much in common. All are working with information as their raw material, and are therefore surrounded by equipment for information processing, such as the computer, fax machine and telephone. All designers are also subjected to similar constraints. Whether positioned high or low on the decision-making hierarchy the designer is given a framework within which to work by those further up the hierarchy, and makes decisions which act as the framework for those lower down. In addition all designers are influenced by the society in which they live, and their decisions must be guided by political, social and financial pressures. These pressures may be due to the spontaneous (or orchestrated) expression of public opinion, or may be formalised in legislation.

Chapter 2
Design as innovation

Introduction

Engineering design has an important role to play in the cycle of technological change. It may be instigated in response to a variety of economic, social and political needs and can involve individuals and both commercial and non-commercial organisations. A linking factor between these is the innovation process. Whether an organisation is a commercial business, aiming to maintain or improve its competitive position, or is non-commercial, but still aiming to be more effective, it must constantly seek to innovate. Innovative activity may take a number of different forms, such as introducing new products or processes, developing improved services, restructuring a company, training personnel in new skills and techniques, and so on. When this leads to the creation of new or improved products or production processes, or the adaptation of these to satisfy new needs, then it is likely that some form of design activity will be required. In this chapter four perspectives are taken of engineering design as an innovative activity. First, the different forms of innovation will be considered. Second, having highlighted the central role which engineering design may fulfil in this, the different types of engineering design activity will be examined. Third the major risk factors associated with design projects will be addressed, along with the constraints which may be placed on the designer as a consequence. Finally, as a large proportion of design activity occurs within commercial organisations, the relationships between the different types of design activity and business performance will be examined.

The scope of innovation

Innovation may be defined from a number of different perspectives. A useful and simple definition, which is broad in scope, is 'the successful exploitation of new ideas' (HMSO, 1994). Although this is a simple definition, it does express three important aspects of innovation. First, innovation is concerned with new ideas. These may take many different forms, both technical and non-technical in nature; they may originate from many different sources; and they may imply varying degrees of change. Some general examples are: ideas for a new or improved product or service, a more effective way of performing a production process, or a new economic policy. Additionally, to be classed as an innovation, an idea must be both realised and able to meet any technical, social or economic criteria which may be used for its justification.

In the previous chapter, engineering design was described as a process for satisfying perceived needs through the creation of technical solutions to problems. It was also shown that design, through the cycle of technological change, may also create the possibility of new sources of needs. Innovation may be associated either with the use (or adoption) of technology, or with the creation of technology. Buying a new machine tool to perform a production process is an example of the former, but it will be apparent that engineering design, through the development of new technology or adaptation of existing technology, is mainly concerned with the latter. It involves access to scientific principles, knowledge and experience, information, techniques and skills, which are brought together in a creative way. Within this more specific perspective of innovation, an expanded definition is 'a process which covers the use of knowledge or relevant information for the creation and introduction of something that is new and useful' (Holt, 1983).

The innovation process is one way that a business can derive a source of competitive advantage, allowing it to maintain or increase its share of markets, sales revenue, and profitability in relation to a particular product. Competitive advantage may be derived through price competitiveness. It is also possible to distinguish the product from the competition through the product's features, its performance, the technology employed, and any support services provided. Support services include after-sales servicing and maintenance, spares sales, technical support and advice. These service factors are distinct from the product's design, and will therefore not be addressed further. Another means by which competitive advantage may be derived is through the production process actually employed in making the product. This may be due to benefits gained directly from improvements relating to the cost or quality of the production process. Alternatively, possibilities may arise for product innovations as a consequence of process innovations themselves.

There have been numerous attempts to categorise the different types of innovation. A useful schema for classifying and capturing the scope of innovation is to distinguish between technological innovation and social innovation (after Braun, 1980. In order to avoid confusion in the text, Braun's use of the term 'technical innovation' has been replaced with 'technological innovation'). These two generic types will be examined in some detail below.

The first category referred to is technological innovation. By focusing on the outcomes or effects of innovation, an important distinction can be made between the viewpoints of the developer (i.e. manufacturer) and the adopter (i.e. customer) of an innovation. Using these two dimensions, a matrix can be devised with four types of innovative activity: incremental, technical, application and radical (after Gobeli and Brown, 1987). This is shown in Figure 2.1. Incremental and radical innovations represent the two extremes of the spectrum of innovative activity. Incremental innovation typically involves only low or moderate advances in technology and is focused on existing markets. (For example the conventional electric kettle has been improved by incorporating the electric element into its base, so enabling smaller quantities of water to be boiled.) Conversely, radical innovation involves both substantial technical developments and the entry into new markets (e.g. microwave ovens). Concentrating on existing markets while introducing significant changes in technology, through a major product enhancement or a new product, is referred to as technical innovation. (An example of this from the recording industry is the introduction of compact disks in competition with tapes and vinyl records.) A relatively low degree of technical change, such as when

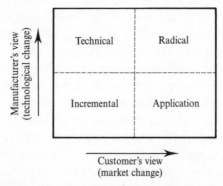

Figure 2.1   Product innovation matrix and the four types of innovation activity.

existing products are modified, may enable the firm to enter new markets. This form of activity is known as application innovation (such as extending the use of compact disks to act as a storage medium for computers, and now referred to as CD ROM).

The second category referred to is social innovation. Whereas technological innovations are concerned with the product or process outcomes of the innovation process, social innovations are important to their successful introduction. Social innovations may lead to improvements in the effectiveness and efficiency of the innovative processes of the business. Such improvements may involve innovations in the organisation and/or people. A business may improve its management by effecting changes to its organisational structures, strategy, personnel resources, methods and procedures, and so on. The effectiveness of personnel may be enhanced through such measures as education and training (skills), and recruitment. These issues clearly influence the design environment, but also affect the designer personally.

From the foregoing, it can be appreciated that engineering design is of central importance to the innovation process. It is a key means by which the desired product, process and market outcomes of technological innovation are realised, and its effectiveness is strongly influenced by the social aspects of innovation described above. But exactly how do the two concepts of design and innovation relate? It is easy to confuse the two and slip from one use of these terms to the other. In general, engineering design is only a part of the innovation process. It is that part of the innovation process that can start with a brief and notionally end with a product ready for manufacture. It can include the development of design concepts and prototypes, detailed drawings, technical specifications and other instructions required for manufacture of the product. However, as will be discussed later in the book, the scope of engineering design can extend both upstream and downstream of these activities.

Invention and creativity are terms often used in association with both design and innovation. Yet their respective meanings and usage are often confused. Referring to the definitions above, it is evident that in the process of innovation the innovator may actually engage in invention. Invention, however, does not necessarily imply the creation of something which is regarded as being both useful and successful. Hence the outcome may not be an innovation. Creativity, on the other hand, has been described as a thinking process which results in the production of ideas that are both novel and worthwhile (Holt, 1983). As such, creativity is an important feature in the processes of design, invention and innovation.

Types of design activity

There are numerous definitions of design. However, it is generally recognised in the field of design theory that three distinct types of design activity exist. These were defined by Pahl and Beitz (1984) in their influential text on engineering design.

- Original design – this involves elaborating on an original solution principle for a system which performs a task which may be new, or may have been solved previously by other means.

- Adaptive design – this involves adapting a known system to a changed task, the solution principle remaining the same.

- Variant design – this involves varying the size and/or arrangement of certain aspects of the chosen system, the function and solution principle remaining the same.

In terms of actual effort required to generate a solution, original design activity is obviously the most onerous, followed by adaptive and variant design respectively. Consider, for example, the requirement to design a system which could be used to transport and locate a satellite in orbit around the earth and return safely with its crew to earth so that it can be used again. The known solution to this problem, the Space Shuttle, took many years of team effort and expertise. It is easily appreciated that moving from identification of the need to the specification of a favoured solution was exceptionally time consuming.

Consider now, by way of contrast, examples of adaptive and variant design activity. A good example of adaptive design is the keyboard as an input device for computers. The mechanical typewriter as a device for transforming information into a documented format had existed for many years prior to the advent of computers. From this the solution principle of the keyboard could be easily developed to suit the requirements of the computer. Finally, an example of variant design is the container ship to transport unified cargo. The solution principle itself was devised approximately thirty years ago, and all container ship designs today are parametric variations for the same standard-sized unit.

Interestingly, adaptive and variant design activities are estimated to account for an overwhelming proportion of all engineering design activity. Some of the reasons for this will be elucidated in the following sections.

The risks of innovating

When organisations engage in innovation they expose themselves to various forms of risk. The greater the degree to which an organisation is prepared to innovate, the greater the potential risks. Earlier it was stated that technological innovation may be related to changes in either technology or markets. With technological change it is also possible to distinguish between product and process developments. The three primary means for technological innovation, and therefore sources of risk, relate to the market, the product and the production process.

As far as risk relating to the market is concerned, the introduction of new or improved products into an existing market, with which the business is familiar, represents a low-risk strategy. Conversely, launching products into new markets of which a business has limited knowledge, and possibly lacks the necessary marketing capabilities, inevitably results in higher risk. Furthermore, in some industrial markets, customers themselves may actually be adverse to new technology, and it is therefore important to understand the market's willingness to adopt the new innovation.

Innovation in the product can take many forms. This may range from the incorporation of incremental design changes associated with the need for improved product features, through enhancements associated with increased performance or functionality, to completely new products which incorporate new technologies. Some of these latter types of design project may need to be supported by additional research and development. They inherently involve more risk, with the possibility of technical failure, cost over-runs and time delays, or even cancellation after considerable resource expenditure. Many large engineering projects, with their particular scope and scale problems for design, serve to remind us of this. The Channel Tunnel project, completed in 1994, is a good example. For many years the British and French governments had considered, but deferred, constructing a tunnel under the English Channel. In 1981 the two governments agreed to construct a tunnel, which would form the basis for a rail link between London and Paris for freight and passenger services. This was a project of immense scale. The opening of the tunnel and start of services were originally planned for 1993. However, there were delays in the construction of the tunnel and technical problems were encountered during the introduction of the service. As a result it entered limited service in mid-1994, with full services progressively established through to late 1995, and at a cost substantially increased above that originally estimated. It should be remembered, though, that for every high profile project which has experienced problems, there will be many smaller projects which have similarly encountered setbacks.

Risks with the production process may arise from a number of sources. Innovation in the production process may be a need in itself, or it may occur as a consequence of changes in the product (new materials or a different design configuration) requiring or providing opportunities for new or modified production processes. Process innovation may involve making small changes such as the modification of tooling or the adaptation of existing machinery, through to much larger changes such as the design and acquisition of new equipment and facilities. The risks involved in process innovation are primarily associated with its timely implementation and the successful performance of the new process in terms of its quality, costs and throughput.

The relationship between product innovation, process innovation and risk can be represented in a diagrammatic form as shown in Figure 2.2. The definitions of design activity presented earlier refer to the product design activity. In this representation the three categories of design activity have been extended to include the process design spectrum too.

The three elements of technological innovation described above are not exclusive and innovation in one may result in demands on the other. Changed product features (e.g. functionality) imply technical change, and even the most basic forms of technical change are likely to have implications for the production process. Innovation involves the commitment of resources, sometimes substantial, in areas such as research and development, new production facilities and equipment,

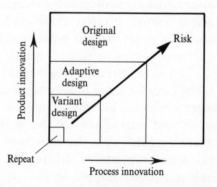

Figure 2.2 Product–process innovation, and the relationship with risk.

product marketing and launch costs, and so on, so that the financial risks may be critical to the viability of the project.

The Channel Tunnel example can be used as an illustration of these risk factors. From a technical perspective, structures such as this have been built before, but it is rare for projects to be on such a large scale. Environmental factors such as the geology of the ground along the planned path of the tunnel, or the possible ingress of water, created uncertainties for the design and subsequent construction. A major source of frustration and delay for the contractors during the project was that the conditions encountered were often more difficult than had been anticipated. Safety was also a particularly important design consideration. During the construction process the safety of individuals had to be considered, and at the very outset many aspects of its operational safety had to be accounted for. From a market perspective, the rail service would be in direct competition with the Channel ferry services. The commercial viability of the project would depend on the operator attracting enough commercial and public customers at a price which would enable sufficient revenue to be generated. At the outset only projections could be made. But to what extent would these be borne out in reality? Some factors would be beyond the control of the tunnel operator, such as how the ferry operators would react to the new competition. Would they cut costs? Would they build new vessels to provide a higher standard of service? Or would they use a combination of measures? From a financial perspective, the project involved the commitment of substantial funds by investors and institutions on the basis of an acceptable return. However, with significant cost and time over-runs, it became necessary to raise additional funds from the major institutions, which would have to accept reduced return on their investments. Also with any project of this nature it is inevitable that there will be political influences. Although the project was financed mainly by private institutions and individuals, decisions made by government, regarding the programme for constructing the new high speed rail link for example, or comments made by politicians, may influence stock market values.

The potential consequences (e.g. economic, social, environmental) of failing to deal with these risk factors means that it may be appropriate for the decision makers who establish the project's framework to place constraints upon the designer, as discussed in the previous chapter. One example of this is the use of legislation, codes of practice, and design rules and regulations. In industries where these are particularly prominent, designers may follow a practice which is sometimes referred to as 'design by rule'. In this case the design requirements, and the scope of the designer's task, are defined in terms of which existing and new technologies, standards and regulations, are to be used. There are sound commercial reasons for why such constraints are applied, and these will be considered below in the discussion of innovation strategies. But because companies expend considerable resources, over time, in developing their knowledge and technologies, they have a legitimate interest in seeking a commercial return on these, either through the improvement of rules, or through the adoption of new and related technologies into the rules.

These issues help explain the reasoning behind the fact that the vast majority of design is of a non-original character. However, although in many situations the designer is constrained, opportunities may still arise within the framework of the project which permit a creative direction to be taken. It is therefore important that designers be encouraged to seek creative solutions to problems, even for the apparently more mundane design tasks. In this way, there will always be opportunities for the designer to derive significant personal satisfaction. With the inherently complex nature of most design work, the role of engineering designers, both individually and collectively within a team, may therefore be characterised by substantial self-discipline on the one hand, and the ability to be creative on the other.

Innovation strategies

Every organisation is confronted with a range of strategic opportunities and has to select a strategy based on its specific circumstances and established goals. Earlier in this chapter, the product innovation matrix was introduced and four basic types of innovative activity were identified – incremental, technical, application and radical. It may be apparent that each type of activity infers different constraints and implications for the designer. This includes the extent to which new concepts and technologies may be used, the relative emphasis on either the product or process design, and designing for different market requirements. In addition the necessary accompanying design disciplines may include technical or functional design, creative strategies, standards, computer-aided design, cost reduction methods, and so on. It was also stated at the outset of this chapter that businesses engage in product innovation in order to increase their competitive position within their chosen markets, and ultimately to improve their economic performance. Thus, an important question for businesses is whether the form of innovative activity they pursue will impact on their likely performance, and if so, what is the most appropriate innovation strategy to adopt.

Research has shown that there is a strong connection between the product strategy a company chooses to adopt and the performance it achieves. One source (Cooper, 1984) has identified five new product strategies, each associated with different performance characteristics. Of these, a strategy featuring a balance of technological orientation with a strong market focus, and which targeted high potential growth markets, resulted in by far the best performance. Similarly, among a sample of small technology-based firms (Meyer and Roberts, 1986), those that over the course of their evolution primarily focused on developing their core technologies for application in familiar markets tended to outperform those that did not. It was also identified that those firms that developed products for new market areas were most successful if they directed their efforts on their existing technology.

It follows that firms which employ some restriction in the degree of technological and market change (indicated by the shaded area in Figure 2.3) will generally outperform companies with a wide diversity in their successive products. This still leaves a wide scope in the choice of product strategy. The available evidence suggests that, all other things being equal, certain strategies will be generally more successful than others. It is important to realise, however, that each of the four basic types of innovation will have relevance in particular circumstances. However, the main concern here is not with the product strategy itself, but rather to

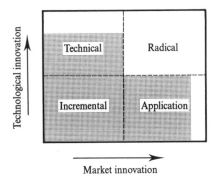

Figure 2.3 Product innovation: strategies and performance.
Most consistently successful strategies for improving a company's performance
are shaded.

recognise that decision makers have a legitimate interest in adopting a particular strategy. This in turn will have implications for the practice of design. An example will help to illustrate this.

The development of a new generation of personal computer which, through developments in core technologies (processor, memory, data storage, and operating system), provides the user with a substantial increase in performance over existing products, would be regarded as a technical innovation. This might require the following: the input of specialist research and development resources between different companies and at different locations; the development and evaluation of different technical concepts; a thorough programme of prototyping and performance testing; and a detailed consideration of the design of the manufacturing system. Alternatively, the predecessor to the new product may have been aimed at the business user. With some adaptive design to improve its appearance and to reduce its manufacturing cost, it may be decided that it could be sold into domestic or small business markets. A choice of preloaded software could be provided as a differentiating feature, such as business accounts, word processing or educational software. This would be regarded as an application innovation. Here it is likely that the design activity will not be so concerned with technical performance, but on cost reduction. The product could therefore be redesigned to improve its manufacturability. This could be achieved by reducing the number of assemblies and introducing cheaper components of a lower standard of reliability, provided that this is deemed acceptable to the market.

The Dyson Dual Cyclone vacuum cleaner: a case study

Vacuum cleaners have worked on the same principle since they were invented in 1901. A vacuum is created around a paper or cloth bag, and the higher atmospheric pressure outside draws into the bag air carrying dust and dirt particles. The air is then expelled through minute pores in the surface material of the bag, whilst the dust and dirt particles (which are too large to pass through) become trapped inside. Dyson Appliances Ltd, a small UK company founded by the British inventor and innovator James Dyson, has developed a vacuum cleaner which has eliminated the need for a bag, by using a radically different cyclonic technology. The story behind what has proved to be a successful and revolutionary product is as much about the inventor's struggle to get the product onto the market as it is about the product itself.

In 1979, James Dyson's idea for a new vacuum cleaner was inspired by the solution to a manufacturing problem with another of his products. The air filter in a spray finishing room was repeatedly clogging up with powder particles. To overcome this, he designed and built a device which removed the powder particles by centrifugal force. Recognising that a vacuum cleaner bag clogs up with dust in a similar way as that in which the air filter had, he realised that the same principle might work in a vacuum cleaner. A five-year design and development process then followed.

Variations of the cyclone design were licensed to companies in Japan, the USA, and Italy. However, a major milestone for Dyson occurred in 1987 when a large American corporation ignored Dyson's patents and began manufacturing the invention, assuming he would not risk an expensive legal battle. It took five years before the American courts acknowledged that the patents had been infringed, and ruled in his favour.

Dyson then focused on his plan to design and manufacture a new model in the UK, convinced that he could design a better product than the major competitors were offering. In 1993, by borrowing nearly half a million pounds from the bank and selling the Japanese rights to the cyclone design to his former licensee, Dyson opened his own research centre and factory, and developed an improved machine that could collect even finer particles of dust. The result was the Dyson Dual Cyclone™ vacuum cleaner (Plate 2.1).

The principle of the Dual Cyclone is shown in Plate 2.2. Air containing dust and dirt enters the outer cyclone at a tangent and spins at speeds of approximately 90 metres per second, so that all large debris and most of the finer dust is forced to the sides of the bin and then collected at the bottom. The air then passes from the outer cyclone through a shroud and enters the cone-shaped inner cyclone, where it spirals downwards reaching speeds of about 400 metres per second, creating powerful inertial forces which spin out the smallest particles (including the microscopic particles that cause allergies). It is claimed that, unlike conventional cleaners in which dust and dirt progressively block the pores and so reduce the air flow and suction, Dyson's cleaner maintains maximum suction at all times as there is nothing to block the air flow.

Dyson's design has enabled the company to challenge the market dominance of the major American and Japanese manufacturers. Indeed,

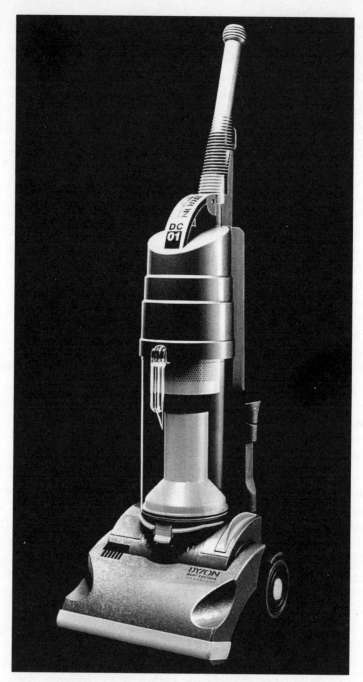

Plate 2.1　The Dyson Dual Cyclone vacuum cleaner
(courtesy of Dyson Appliances Ltd).

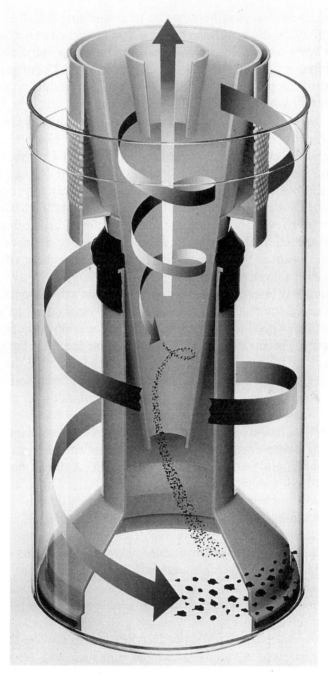

Plate 2.2   The operating principle of the Dyson Dual Cyclone vacuum cleaner
(courtesy of Dyson Appliances Ltd).

by the end of 1995 Dyson became the largest selling brand in the UK market, outselling the well-known and long-established manufacturers.

There were several factors which contributed to the Dyson Dual Cyclone cleaner reaching the market successfully. These provide an illustration of the positive impacts of good product design as well as the possible pitfalls which await the designer/innovator.

When James Dyson approached the major manufacturers of vacuum cleaners with his idea they turned him down. They chose to invest in research and development of their existing bag-based technology (i.e. incremental innovation). With the benefit of hindsight one can question why they did not buy up Dyson's patent rights, if only to prevent their exploitation. With so much invested in conventional cleaners, and significant sales revenues being generated by sales of replacement bags, it can be argued that they missed an opportunity to protect their competitive position by stopping the further development of Dyson's cyclonic concept there and then.

A particularly interesting feature of the product's development is how Dyson managed the risks involved. Although Dyson's product was very innovative (a new product and new market for the company), no stage of its development involved truly radical innovation (i.e. developing a new core technology for a new market). He had already applied the cyclonic technology elsewhere and although some further development of the cyclonic technology was involved, it was predominantly an example of application innovation (i.e. adapting an existing technology to a new application). Moreover, the exposure to commercial risk during the early development stages was reduced by licensing the designs to other manufacturers, while subsequent development of the Dual Cyclone design involved several stages of adaptive design (i.e. incremental innovation of the company's core technology) which inherently involves less risk. Dyson has also introduced the technology to other markets, as he has adapted the cyclonic principle to develop a filter for the silencer of a diesel vehicle's exhaust system. Indeed, Dyson's innovation strategy is similar to those of many other successful small companies discussed earlier in this chapter – focusing on and developing its core technologies for existing markets, or adapting these to create alternative applications in other markets.

Dyson demonstrates a difficulty faced by many smaller companies. Although they may be innovative, opportunistic, and responsive to market changes and new ideas, they are unlikely to have the resources to undertake substantial technical innovation on the same scale as larger companies. In contrast, if one of the major manufacturers had taken up the technology at the outset, they may have been able to develop it much more quickly, and to introduce it simultaneously in many of the global markets.

From a design and marketing perspective, Dyson emphasised the quality of his product. His policy was to go for a product that was technologically better, but which also had an exciting design (it combined a high-tech styling with unusual yellow and grey colours) that was distinctive when set against the competitor's products, whose subtle differences made their looks almost indistinguishable. This was important to a small growing company like Dyson launching a new product. A company with a high market share and established distribution can go for products with a low profit margin and high sales volume, but Dyson, starting from a small base, needed to be able to make a profit on low initial sales. The product's distinctive features enabled a premium price to be obtained consistent with a reasonable sales volume.

The product's distinctive characteristics also meant that, with only two models – an upright (shown in Plate 2.1) and subsequently a cylinder model – it was clearly differentiated from the competitors' products. In comparison, as all the competitors' products were very similar they differentiated their products by manufacturing a wide product range, with models offering different features to alternative market segments.

The long-term enthusiasm and commitment of Dyson was also an important factor. Apart from the missionary-like belief in his product, developing the business inevitably required finance to be raised. In order to maintain personal control he had to dispose of some of his other businesses. Despite its high cost, taking out world-wide patents on his designs proved to be a wise decision. These enabled him to seek redress when another company sought to manufacture his design. Fortunately he had the necessary resources to defend the breach of his patent.

The Dyson Dual Cyclone cleaner provides an example of good product design and the important role that design has in creating successful products and businesses. It also illustrates the importance of enthusiasm and commitment, and provides insights into the management of the risks involved in product innovation.

Summary

From the preceding discussions it is apparent that due to a number of factors which arise out of the environment of design, there may be sound reasons for those decision makers empowered with the responsibility for establishing a project's framework to place some constraints upon the designer. The social, technical and economic factors are of particular significance. In this chapter two specific aspects of this have been examined. These relate to the sources of risk (and the necessary contingencies required to manage it) and to the implications arising from the type of innovation strategy adopted by a business. Having recognised the presence of these factors, it is apparent that in most instances it will also be in the company's own interest to place some form of constraint upon the designer. The imposition of such constraints has inevitable consequences for the characteristics of the engineering design activity, and these will be dealt with in subsequent parts of the book.

Chapter 3
The role of the designer

Introduction

There is no such thing as a typical designer or, for that matter, a typical designer's job description. However, given that designers form an identifiable professional group it is possible to describe their responsibilities and expertise, and to relate these to the responsibilities and expertise of other groups involved in the process of developing products. The objective of this chapter is to provide such a description, and in particular to provide an insight into designers' use of existing technologies to meet specified needs.

From design brief to documentation

In the previous chapters, we described how design projects are circumscribed by decision-making processes which are not regarded as being part of the activity of designing. As a result of these external decision-making processes, the proposed product is always partially defined – in terms of its function at least, if not its embodiment as well – before designers start to work on it. When designers start their work, they need to be made aware of the initial problem definition so that they can focus their attention on the range of design solutions that may satisfy it. They also need to know what criteria they should apply in order that they can identify the 'best' solutions. Finally, they need to be told what sort of information they are expected to generate. Descriptions and sketches of two or three design concepts may be all that is required for one project, whereas another project may involve producing detailed documentation of performance calculations in addition to manufacturing and assembly drawings, production schedules, and cost data.

Taken together, the information which is given to designers at the start of a design project is often referred to as the *design brief*. It may be given to them either verbally or as a written document. The amount of information involved can vary enormously from one design brief to another – designers can be presented with anything from a vague idea on a single-page fax message to a document with hundreds of pages. However minimal or extensive it is, though, the design brief forms the framework for the designers' subsequent work, as it tells them what is to be designed and, for that design, just what level of detail is required.

Although the design brief may include requirements, or at least assumptions, relating to the embodiment of the product, its primary concern is to specify the product's functionality. It is the designer's job to decide how best to implement this specified functionality, and to present a design solution in the form of drawings and documentation. Just as the design brief is the starting point for the designer's activities in connection with the product, the completed drawings and documentation mark the end of these activities.

Activities of the designer

As the outcome of the designer's activities consists of final drawings and documentation, these activities must clearly include writing and drawing – whether on paper or using computer-based document processing and draughting tools. The final documents, however, are merely a means of describing one or more particular designs. It is the formulation of these designs which is the essence of the designer's role, involving them in a range of activities before the final drawings and documentation can even begin to be produced. The nature and ordering of these activities has been the subject of a great deal of attention from design theorists. It is not the purpose of this chapter to enter too deeply into this complex subject, which will be dealt with in detail in Section 2, but the designer's role cannot be adequately described without making some preliminary observations.

As has already been noted, the designer's task in relation to a particular product is to produce a description of some realisable technological hardware (or software) to meet the given functional specification. In other words, given a specification of what the product should do, designers have to provide a description of what it could be.

To do this, designers engage themselves in three broad types of activity (Cross, 1989). First, they generate ideas for possible design solutions. Second, they evaluate their alternative solutions with regard to the specified requirements and goals. Finally, having selected the most appropriate solution, they communicate design intent to other people involved with the product, using drawings and other means to transfer information about their design solution.

Generating ideas for possible design solutions is not a straightforward process, as anything but the simplest product will be composed of a number of systems, which in turn may be composed of sub-systems. Certain strategies, which will be discussed later in this book, can be used to construct this assemblage of systems and sub-systems. However, designers must also be able to draw on their knowledge of the relevant technologies, together with information gathered from books, suppliers' catalogues and other sources. Without such knowledge and information, designers would be incapable of proposing a solution to even the simplest design problem. The generation of design solutions is rarely carried out in isolation – it is invariably intertwined with the process of evaluating solutions. Information obtained by modelling and evaluating a particular design is often used to inform the generation of an improved design. This feedback process may be repeated as many times as necessary to reach a satisfactory solution.

The activity of modelling, which is discussed fully in the next chapter, is a central one for designers, being the primary means of representing, communicating and evaluating design solutions. Models tell designers whether their solutions will meet the specified requirements, and help them to make sensible choices between alternative design solutions. In some design projects, designers may construct and test 'real' physical models of their designs, or even full-scale prototypes. However, the use of mathematical models is much more common. Mathematical models can be anything from a few scribbled calculations to sophisticated computer-based simulations. Different models may be required in order to investigate different aspects of the performance of design solutions. It is impossible to list all the physical and functional attributes for which specialised modelling techniques are available, but stress and reliability are two examples for which the techniques are highly developed and widely used. The use of computer-based modelling tools has enabled designers to construct accurate models in less time than was previously possible. Another important benefit of computer-based systems is the ease with which data, such as a description of the geometry of a component, can be transferred between different modelling tools. These technological advances have made it possible for designers to evaluate design solutions with a degree of thoroughness and accuracy that would have been inconceivable before the advent of computers.

Finally, the communication of design information is done primarily through the preparation of drawings and other documentation. Indeed, draughting is perhaps the most visible of all the activities designers undertake. However, words are just as important a means of presenting design information as pictures are, so designers may often be required to produce written documents, and even to make verbal presentations of their designs.

Two definitions or 'views' of the design are often required as part of the final documentation. On the one hand, designers are required to provide the information which a potential user of the product might need in order to compare it with other similar products, or to use, maintain and repair it, or even to dispose of it. This is sometimes referred to as a functional definition of the product. On the other hand designers may also be required to specify how the product is to be made. The level of detail required for this manufacturing definition of the product can involve the production of a surprisingly large quantity of information. Material definitions, dimensions, tolerances and surface finishes must be specified to define each component fully. In addition, process plans and assembly instructions are often necessary in order to plan component routing and other production details.

The different areas of activity described above are often carried out by different people, with specialised skills in particular areas. The differentiation between designers and draughtsmen is perhaps the most obvious example of this division of labour in the design team. Given this division, the designers are usually responsible for generating and evaluating design solutions, while the draughtsmen have the task of drawing the selected solution. However, because the designers' definition of the solution is inevitably incomplete, draughtsmen often have to make numerous design decisions in order to 'fill in' much of the detail. Modelling is another area of activity which is often carried out by specialists. In fact, it is not uncommon for different specialists to model and evaluate design solutions with respect to different aspects of their performance, with each one specialising in the use of a particular analytical technique or computer-based modelling tool.

Creativity in design

Design is commonly thought of as a creative process, involving the use of imagination and lateral thinking to create new and different products. This view is somewhat misleading when applied to engineering design. Although creativity and original ideas are often called for in design projects, much design work is not highly original in nature. The majority of engineering design activity falls within the scope of variant design, as defined in the previous chapter.

The predominance of non-original design activity reflects the low level of risk associated with it. Established product types are a safe option for engineering organisations, as both the markets and the technologies involved are familiar and well-understood. As a result, most of their design resources are focused on designing new versions of established product types, whether they are designed to order, or for batch or mass production. This usually leaves the designer with little scope for originality or creativity with regard to the overall design concept but does not necessarily exclude the application of the designer's creativity in the embodiment and detail stages. Indeed, it has been recognised that innovative activity often takes place at the system, sub-system and component levels (Pugh, 1990). New ideas at these levels can contribute significantly to the success of a product, and can be crucial if the product is to be targeted at a new market sector or has to meet a customer's specific requirements.

Although their short-term product design activities can almost always be classified as variant design, engineering organisations often sponsor more free-ranging design activities under their medium- and long-term product development strategies. Thus, designers working in research and development areas often have more freedom and scope for creativity than their colleagues elsewhere in the organisation. Original and adaptive design activity also takes place in other contexts where the degree of risk is less critical than in the context of a commercial organisation. These other contexts range from the garden shed inventor, for whom the investment of time and effort is a pleasure rather than a budgetary cost, through to prestige engineering projects such as the space programme, where the achievement of the goal is deemed to justify the often astronomical expenditures involved. Despite their differences, these two scenarios probably represent the ultimate in creative scope for the designer.

Technical knowledge in design

The technological fabric of our society is composed of innumerable engineering products which help people to cook food, do work, travel to see friends, and so on. Designers play a major part in the creation of this technological habitat. However, in the process of creating technological products, designers must also make use of existing technologies. This is not only through the use of design tools such as drawing boards, pencils, calculators and computer-aided draughting systems. Designers cannot even start to conceptualise solutions to design problems without having some knowledge and understanding of the components from which such solutions might be constructed. At its most fundamental level, this knowledge includes the properties of space and time, the concepts of energy, power and force, and the properties of solids, liquids and gases. However, engineering designers also make use of more specialised components, often relating specifically to their own branch of engineering. For a structural designer, these will include materials such as concrete, steel and wood, as well as structural elements such as columns, arches, ties, struts and beams. An electronics designer will have an understanding of the electrical concepts of current and potential, and will be able to make use of components such as conductors, resistors, capacitors, and transistors.

Designers often make use of larger 'building blocks' than those mentioned above. In designing a complex structure, the structural designer may choose to incorporate a number of trusses which can be supplied, prefabricated, by a specialist firm. Despite the complexity of their truss designs, the amount of data provided in the firm's catalogue is relatively small. The data given for each design consists of a few overall parameters such as dimensions, weight, stiffness and load-bearing capacity. The catalogue also contains details of the fixing points where the trusses will be attached to other structural components. These are the essential pieces of information which the designer needs to select an appropriate design and to build it into the overall design of the structure. The designer need not be concerned with the details of the truss design, its geometry, the selection of the component elements and the method of fabrication. The truss can be treated as a 'black box' with specified characteristics and interface points, as shown in Figure 3.1.

Thus, the use of a predesigned component enables the designer to avoid some detailed design work. It also eliminates uncertainty from an area of the overall design, as the specification and cost of the component

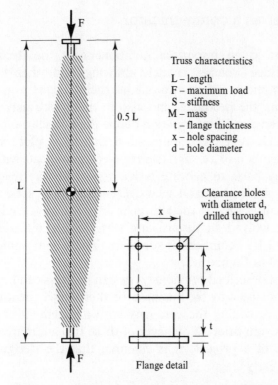

Truss characteristics

L – length
F – maximum load
S – stiffness
M – mass
t – flange thickness
x – hole spacing
d – hole diameter

Clearance holes
with diameter d,
drilled through

Flange detail

Figure 3.1 Standard components: a structural truss.
A designer wishing to make use of a predesigned component as part of a larger system does not need to know and understand all of its inner workings. In the case of this structural truss, the overall characteristics of the truss can be defined without reference to its internal geometry. Precise geometric descriptions need only be provided for the flanges which form the interfaces with other structural components.

in question is known. Someone else has already designed it, and someone is undertaking to make and supply it.

Trade magazines and professional journals carry numerous advertisements for products which are specifically designed to be designed into other, larger products. Such advertisements inform designers of the existence of these products. However, in order to be able to take advantage of such components, designers not only need to be aware that they exist but also need good, reliable information about them. Full technical specifications of these products are usually provided in suppliers' catalogues. These are a crucial source of information for designers, and are often found alongside engineering textbooks and other handbooks on designers' bookshelves.

The designer as a communicator

Communication is an important component of the designer's role. Designers produce sketches, models, drawings and other documents in order to communicate information about their designs to those responsible for making the products, and also to those who may wish to use them. These transfers of information define the boundaries of the design process itself. However, communication also takes place within these boundaries. Members of a design team must communicate with each other on an everyday basis to allocate tasks, report on the results of their activities, and share ideas and views. Designers may seek advice from technical specialists such as metallurgists or production engineers. They may also be involved in discussions with people outside their own organisation, such as component suppliers. These communication paths are summarised in Figure 3.2.

As has been described previously, design projects can begin with the identification of a need by the business, or alternatively be initiated by an external client providing the company with a statement of its requirements in the design brief. Whichever of these two alternatives results in the instigation of a project, it is essential that the designer becomes involved at an early stage.

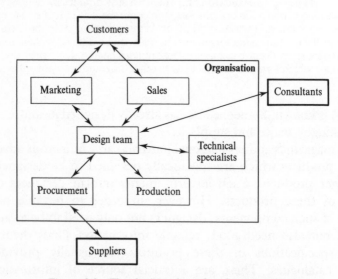

Figure 3.2 The designer's paths of communication.
To assemble the information they need, designers must maintain communications paths with other functional groups both within the organisation and outside it.

Consider first the case of a company which is provided with a statement of requirements. The first task that the designer must undertake is to clarify with the client the scope of design activity. Negotiations should establish what the company is actually expected to deliver (i.e. the extent of supply in terms of drawings, documentation, product, after-sales service and so forth), the dates when the company is expected to deliver its outputs to the client, any contractual contingencies (e.g. penalty clauses that will apply in the event of missed delivery deadlines) and the price to be paid for the work that the company is undertaking. As well as these aspects, the designer may need to clarify the technical requirements with the client, as this will be used as the basis for all subsequent design work undertaken by the company. In the case of the bridge example (which was explored in Chapter 1) the designer needs to establish exactly what static loading the bridge has to withstand, the span to be covered and any additional environmental loading conditions that will have an impact on the design of the product. It is important that the designer is able to listen to the client, interpret the client's wishes and develop these into an agreed definition of the technical requirements. For major contracts, and possibly for many minor projects, this clarification process can take a significant period of time, but the time is justifiable as mistakes made at this early stage of the process can prove to be expensive. A little more effort at this stage may prevent major problems later in the design process.

Now consider the case of a company which identifies a new market need itself. Once the marketing department within the company has pinpointed a new business opportunity, the designer undertakes an assessment of the company's ability to generate a suitable product. It is necessary therefore that the designer engages in consultation with representatives from marketing in order to develop a full understanding of the nature of the opportunity (i.e. in terms of the technical requirements, the potential sales volume, and the impact on the workload of the business) and evaluate the level of risk to the business of embarking on the venture. As with the first example, the result of these communications is an agreed set of requirements and once again it is important that this brief contains sufficient information to enable a good start to be made to the design process.

Once the technical requirements have been defined and the company has been given the go-ahead for an order, the designer must make contact with suppliers. Many engineering companies do not retain in-house all the necessary production capabilities that they require to produce all the parts that might constitute a product. Indeed, in today's very competitive markets, it makes economic sense to contract out some of the piece-part manufacturing. Companies are able to realise benefits from establishing healthy long-term relationships with suppliers of key components. However, the development of partnerships between companies and suppliers over recent years has resulted in designers being engaged in an increasing amount of discussion and consultation. Of course, the designer's concern in such discussions is primarily related to the product's functional performance and quality, although other issues cannot be ignored. Whether the design is done by the company, the supplier, or by both in collaboration, the designer has to communicate exactly what it is that the supplier is expected to produce. Traditionally, the designer has communicated this design intent through drawings and supporting documentation, but with the increasing sophistication of computer-aided design and manufacture software, designers are now able to transfer drawings electronically and so make the process of communication appear almost seamless.

After the designer has identified satisfactory suppliers to provide all those parts of the product that the company has decided not to manufacture in-house, the process of preparing the manufacturing and assembly instructions for internal departments and suppliers can commence. To achieve this the designer must communicate with internal manufacturing personnel in order to develop a detailed understanding of the capability of the production facility. Over recent years it has become increasingly important for designers to consider all aspects in producing a design. Issues such as design for manufacturability, design for assembly, and design for quality all impact on the ability of the company to extract profit from the project. Additionally, it is important that the designer is available throughout the manufacturing and assembly process to troubleshoot any problems that might arise. Ongoing communication and collaboration with production personnel is crucial to ensure smooth and efficient flow from concept drawing to manufacture of the product.

Finally, it should be realised that the communication between the designer and client (or marketing personnel) is an ongoing process. As mentioned earlier this exchange begins with the clarification of the brief, but this is by no means the only time communication between the two parties takes place. Indeed, the communication continues throughout the entire period of the project, and may extend into after-sales technical assistance. For example, once the designer has decided on the overall nature of the concept solution an opportunity often exists to engage in some form of dialogue to establish whether the client is satisfied with the proposals (i.e. at the end of the generation stage of design activity). In this way any changes that need to be reflected in the preferred concept solution can then be incorporated prior to the company starting detailed design work. At the other end of the process, if some difficulty is encountered in operating the product the designer should be available to provide technical guidance to alleviate the problems.

Thus, as the discussion above has shown, in the majority of practical engineering design situations a designer does not operate in isolation. Apart from the obvious requirement to work and communicate with fellow designers in the team, there is a whole host of other people who must be consulted in order that a design actually makes the successful transition from drawing board to product.

The designer as a professional engineer

Engineering designers are part of the engineering profession, and like all professionals have professional obligations. Membership of a relevant professional body is expected together with involvement in public debate on all matters pertaining to the professional standing and development of the engineering profession. But professional status also brings legal responsibilities.

It is appropriate to begin with a discussion of the designer as an individual who has a unique combination of talents and concerns. The fact that every single designer is characteristically different means that even when supplied with the same problem, different designers will generate different solutions. Indeed, should several designers be asked to make a selection between a finite range of permissible options, it is likely that each will favour a different solution. The reasons for this are manifold: designers possess different skills; they have distinctly different background experiences; they each apply a different value system to personal and professional matters; and they each have a unique approach to designing. It is clear, then, that the personal attributes and preferences that characterise a designer will have a pervasive influence on the way in which each designer performs their job.

One particular attribute which can have an impact on many of the decisions that a designer takes, or can affect the way in which a designer carries out their work, is that of personal ethics. In the most elementary of examples a designer's ethical stance can prohibit certain design options being taken. It may be the case that an automotive engineer is not content to select a particular catalytic converter because, in comparison with other options, it has a much poorer performance in terms of emissions removal. In a more extreme, but not uncommon, case a designer's ethical stance can actually prohibit certain career options. For example a designer might have strong reservations about working for any company which manufactures armaments. Ethical views, therefore, are an important part of a designer's make-up and their potential impact always has to be taken into account.

Consider now the issues relating to professional obligations. Designers usually begin their professional careers studying for a degree on a suitably certified course at university. After acquiring a degree in an appropriate discipline, however, the designer's professional development is far from complete. Indeed, it could be argued that the degree represents only the first stage of the lengthy learning process that is necessary for a designer to become a fully qualified, competent professional. A crucial part of a designer's ongoing professional development is gaining the breadth of experience necessary for chartered engineer status, as this is a requirement for many senior positions. To achieve this a designer must necessarily be associated with an appropriate institutional body which will eventually recognise the professional and vocational training that has been undertaken. Such professional institutions serve other important purposes as well as offering routes to becoming a chartered engineer. Among these are the opportunities that institutions offer to members for further learning and development, as all such institutions publish learned transactions, hold regular conferences and seminars, and publish informative journals which are of immense use to the designer in maintaining an awareness of current developments. Additionally, such institutions offer the opportunity for designers within a particular discipline to comment on issues of significant professional interest with a unified voice.

One final issue that must be addressed is the legal implications of engaging in design activity. Every product a designer generates is used to a greater or lesser extent in society and, should an accident occur, legal proceedings may result. Often in such cases the burden of responsibility does not fall on the company who manufacture and sell the product; rather, the onus lies with the designer. It is important that designers understand the extent of their responsibility. Product liability is the name of the special branch of law dealing with personal injury, or property or environmental damage resulting from a defect in a product. In certain countries a product liability suit is a common legal action which can lead to three different charges of negligence being levelled at the designer: the product was defectively designed; the design did not include proper safety devices; the designer did not foresee alternative uses of the product (Ullman, 1992).

In the event that any of these charges are brought against a designer, whether founded or not, serious consequences can result. In such circumstances it is not uncommon to find the professional competence of the individual being called into question, and even if the case is found in favour of the designer, the damage done to the professional reputation of that person may stay with them long after the case has finished. When a designer is accused of negligence, the relevant professional institution may be able to provide both legal assistance and technical advice to help disprove the burden of culpability.

Summary

The fundamental role of the designer is to provide a technical description of a product that will satisfy a brief which indicates goals, requirements, and evaluation criteria. The designer achieves this by creatively using a broad range of knowledge, of both a general and a domain-specific nature. In the process, communication is necessary with the many other parties also involved in the joint task of converting an abstract idea into a concrete reality. In some cases the designer's role is a pivotal one, orchestrating the many participants and integrating a variety of conflicting demands. In this activity the designer must not ignore the responsibilities associated with professional status. All decisions must be made, therefore, with due regard to both legal requirements and ethical standards.

Section II
Theories of
Engineering Design

Chapter 4
From craft to CAD–CAM

Introduction

Treating design as an independent activity, which is separated from the manufacturing process, is not the only way in which artefacts can be created. Indeed, in the past, design as an identifiable activity did not exist. Take for example the design and manufacture of a small wooden fishing boat. The fisherman would approach a boat-builder whose reputation was good and whose work he respected, and during a brief discussion would outline the functionality he required to that boat-builder. Between the two of them they would decide on the overall dimensions (and agree a price) and then, with no more detailed instructions but the implicit understanding that the vessel would be similar to the craftsman's previous vessels, the boat-builder would set to and produce the new craft. In this particular case it could be said that the design consisted of the two or three templates used by the builder for every vessel he built. This craft process contrasts markedly with the image conjured up by *CAD–CAM*, the acronym for computer-aided design and computer-aided manufacture. The ideal scenario for this process is that all aspects of design and manufacture are directly controlled by the designers. In an interactive process between humans and computers an electronic description of the product is built up. This can be displayed on a high resolution colour monitor, and tested and evaluated using appropriate software, until the designers are satisfied. On completion this description is sent electronically to manufacturing where computer-controlled tools produce the product automatically, with little or no further human intervention.

Clearly these two situations describe the extremes of a spectrum of possibilities, and although it appears that one belongs to a distant past, and the other to a fanciful future, it is possible to find examples of both practices being carried out today. In many industries, however, the latter practice is evolving on the basis of the former. Before discussing how and why the processes of design and manufacture develop from craft-based procedures to CAD–CAM, it is worth examining each in a little more detail.

In previous centuries most industry was craft-based. The very concept of craftsmanship embodies the idea that the quality of the work is the result of experience and knowledge, not of careful design and drawings. In the craft-based industries of the past, such as pottery, tailoring, boat building, coach building and so on, the manufacturer was responsible for the design, and in many cases appeared to design the product as it was being created. It could be argued that a rough plan existed in the form of patterns, or at least principal dimensions, but in some cases it was just the memory of how it should be done, an imperative which embodied the collective experience of generations of earlier craftsmen. It is certainly true that all the detailed decisions were made during the manufacturing process. In fact both the overall concept and the details of each artefact were not created by any individual, but evolved over many years. The failure of a particular feature would necessitate modification to subsequent versions, while successful features would be retained and refined. Deciding on the specific dimensions of a component would now be considered the function of the designer. In a craft-based industry these decisions were all made by the craftsmen themselves, looking to tradition and past practice for guidance.

Today, in certain industries, the relative importance of the two principal roles has been reversed. In these cases it is the manufacturer who is all but invisible, while the designer is able to reach so far into the manufacturing process that the product can be realised without further intervention from others. One of the first examples of this transfer of skill from the shop floor to the design office was in the printing industry, where the tasks of the typesetter, graphic designer, and even editor could all be accomplished by the author or journalist alone. It is easy to see how similar technology can also allow the designer of printed circuit boards to directly control their manufacture as well. Clearly it is computers that have enabled this transfer of power, and as the control of more manufacturing equipment becomes automated, so this type of production will become more widespread. In mechanical design it is now possible to sit at a computer workstation, sketch in a proposed component, analyse and refine it until a satisfactory definition of the preferred solution is achieved, and then, by directly linking the computer to equipment such as sintering machines (where lasers fuse layers of dust into the specified shape), quickly create a solid model of the item automatically. While this process of *rapid prototyping* is too expensive to use for the manufacture of delivered parts, and indeed would be an inappropriate procedure for mass production, it can be used to produce prototypes or full size facsimiles for direct assessment at the earliest stages of design.

The growth of design

If the two design scenarios described above are extreme cases, then clearly design activity can fall anywhere in the range between them. Historically the development of design has been a gradual progression from the minimal contribution found in craft industries toward the design-dominated procedures epitomised by CAD–CAM. This progression describes a gradual growth in the contribution and importance of design.

The traditional skill of the designer is draughtsmanship, and the ability of the designer to draw the product before the manufacturer built it was the original contribution of the design process. The designer's first concern is therefore with geometry, for the drawing defines the shape of the product. In creating the drawing the designer can change and modify the geometry to ensure that the dimensions of the design will work. A simple example is that of a passenger railway carriage. It must pass under bridges, and between trackside structures. Clearly it is easier, safer, cheaper and far more efficient for the designer to try out on paper alternative arrangements of wheel diameter, chassis structure, and compartment height than for the builder to produce several versions experimentally on a trial and error basis.

Geometric constraints (which have always been a primary consideration) only dictate some aspects of performance in which designers are interested. If a design of any sort is to work, many other performance considerations have to be taken into account. A simple list for the railway carriage example could include items such as strength, stability, suspension, ventilation, heating and safety. It will be noted that many of these performance aspects are also related to geometry, and by combining measurements taken from the drawing with detailed calculations the performance can be evaluated. For example, the strength and stiffness of a component is dependent on its size and shape and on the characteristics of the material employed. Similarly, stability is described by evaluating moments, and these can be found by calculating forces and locating their point of action on the drawings. Similar considerations apply to many other elements of performance, and this can explain how designers gradually extended their skills to include the many different types of analysis that are required to predict all elements of performance. This aspect of design requires both a broad-based understanding of scientific principles, and a detailed knowledge of domain-specific technologies.

This stage of the growth of design could be considered the midpoint of the spectrum introduced earlier. Ensuring satisfactory performance (and avoiding performance failure) is termed *functional design*: the designer describes what is required, and the manufacturer creates it. But the designer's intimate knowledge of the geometry of the proposed product can be used in another way. Difficulties in producing the design can be anticipated, and modifications made to ensure easy and efficient manufacture. In other words, to functional design can be added *design for manufacture*. Evidently the greater the understanding that the designer has of the particular production process which will be employed, the more effective this aspect of the design will be. For this reason designers were traditionally expected to spend time on the shop floor learning about the limitations of the manufacturing processes, and the corresponding influence these had on any design. In more recent times the importance of ensuring that a comprehensive understanding of the manufacturing process exists within the design office has led to the introduction of alternative procedures. Skilled production workers are often transferred into the design office, bringing their knowledge with them, and suggestions from the shop floor are encouraged (and rewarded) so that production experience continues to be fed into the design process. These are examples of the movement of skill and knowledge 'upstream' in the design and manufacturing process, as it evolves from craft to computer-dominated systems. They are also a manifestation of the transfer into the design office of responsibility for the complete success of the product, including its manufacture.

However, this interpretation of the changing role of the designer does not imply that the manufacturer is being made redundant. Although the designer has to understand the manufacturing procedures (in order to plan how these will be used, and to ensure that the product design takes full advantage of efficient production methods), the manufacturer must still be concerned with developing the production processes themselves. It is the manufacturer's responsibility to improve continually both the efficiency of production and the quality of the product (in terms of confidence that it is being produced accurately to the design specifications).

Indeed, the evolution of manufacturing systems has also seen a gradual transfer of knowledge from the shop floor to the designer, but in this case to the designer of the production process. In the craft industries every worker served an exhaustive apprenticeship lasting from three to seven years, a process that resulted in a highly skilled workforce. Plainly, less training is required for a system where a skilled foreman interprets the drawings and then supervises a team of semi-skilled workers. However, in industries where Henry Ford's ideas on production lines are in operation the tasks may be broken down into such small units of work that each individual on the line can learn, in a matter of hours, everything that is required to perform the tasks (Braverman, 1974). For such a procedure to be successful much of the skill requirement is effectively removed from the worker completely, and built into the machinery or manufacturing system instead. It is therefore the designer of the production line who must have the necessary manufacturing knowledge. Careful design of both the product and the production process can result in sophisticated production lines from which the output is variable (Littler, 1986). This can be achieved by batch manufacturing of parts which, with efficient materials control, can be assembled to each individual customer's order. While it is the automotive industry which leads in this technology, some vehicle manufacturers have further redesigned the production process in a different direction. Groups of workers are gathered together into production cells, and each cell is made responsible for part of the assembly process (Littler, 1985). Such approaches are collectively referred to as *lean production methods*. One of the advantages gained by such systems is that the improved motivation and involvement of the workers results in increased information being passed back into the design office.

Returning to the central design task, that of product design, the increasing scope of the designer's responsibilities does not stop with the addition of the manufacturing definition to the functional one. Following delivery, many products have to be maintained and so design for maintenance is important too. In addition many products will eventually be scrapped and so there is a growing interest in design for recyclability. Plainly, once all these elements of the product life cycle are incorporated as design considerations, there is considerable scope for conflict. For example, the best functional design may not be the ideal for manufacture, or for maintenance, and so on. The increasing range of responsibilities placed on the design function has led to a situation where no one individual has the necessary expertise to consider all aspects, or to resolve the inevitable conflicts. Such design complexity necessitates a team approach, with customers, manufacturers and operators all having valuable contributions to make. If this team is augmented with others involved in marketing, distribution, safety and so on, all aspects of the product can be considered at the outset and catered for in the design. A design operation of this nature is termed *simultaneous* (or *concurrent*) *engineering*.

The increasing importance of the design team reflects the growth in this group's ability to predict the future performance of the design in all stages of its life cycle. The better the predictions as to how something will behave, and how it will have to be made, the better will be the design. Historically the principal tool used for forecasting the product's adequacy has been the drawings, and their representation of the product's geometry. The drawings enable calculations and deductions to be made about a design's characteristics and in effect its future behaviour can be simulated. In this way production difficulties, maintenance problems and performance failures can all be avoided. Simulation of possible future scenarios is the basis of most design evaluation and the growth of design is reflected in the increasing sophistication of simulation techniques.

Simulation and modelling processes

For a designer to predict the future characteristics of a product it must be possible to simulate its behaviour. This can be achieved by examining specific aspects of the design, and by modelling these the behaviour of the final artefact can be predicted. Modelling procedures are an essential element of the design process, for it is through models that the designer is able to establish how the design will be built and to evaluate how it will perform.

Models can take many forms, only one of which is a set of drawings. It has already been suggested that early designers were primarily draughtsmen, and that it was because their drawings modelled the final product that they were so useful. Inspection of such drawings allowed many aspects of the design to be evaluated and the future performance to be predicted. But drawings are only one type of model, and many others are used by the designer. These can be categorised into three types: iconic, analogic, and symbolic (Open University, 1993). We will examine these in turn.

1 *Iconic models* – represent things as they appear, and so sketches and drawings of all types fall into this category, an example being shown in Figure 4.1. Scale models, and full size 'mock-ups' are also of this type.

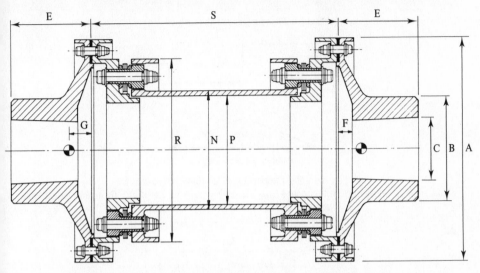

Figure 4.1 An example of an iconic model: a technical drawing.
All sketches and drawings are iconic models. This example is a detailed technical drawing of a flexible coupling which provides accurate information of the design's geometry
(courtesy of Flexibox Ltd).

2 *Analogic models* – represent specific features of the design by using an analogy. Schematic diagrams and flow diagrams are of this type, as are circuit diagrams. Analogic models can also be used to visualise properties that are abstract in nature. For example, shear force and bending moment diagrams, such as those shown in Figure 4.2, are commonplace in structural design. Similarly, diagrams indicating contours of constant value for a given property (such as pressure, temperature, or fluid velocity) are widely used in many disciplines.

3 *Symbolic models* – describe the design, or some aspect of it, by using words, numbers or mathematical symbols. A parts list falls into this category, as does a work schedule, but mathematical equations describing some aspect of the product's behaviour (such as the Rankine–Gordon equation shown in Figure 4.3) are a more sophisticated example of a symbolic model. Computer models are based on complex algorithms made up of many mathematical equations, and therefore are also symbolic

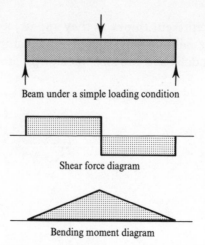

Figure 4.2 Examples of analogic models.
Examples of analogic models are the shear force and bending moment diagrams commonly used in structural design.

$$\sigma_f = \frac{\sigma_y}{1 + (\sigma_y/\pi^2 E)(L^2/k^2)}$$

Figure 4.3 An example of a symbolic model: the Rankine–Gordon formula.
Mathematical equations can be symbolic models of a design. This is the Rankine–Gordon equation which gives the failure stress for structural columns. The formula incorporates Euler's analysis of slender column failure with empirical results to model the interaction of two modes of failure: that of compressive yield, and that of buckling. Given the material properties of yield stress and Young's modulus, and the design variables of column length and radius of gyration of the cross-sectional area, the failure stress of the column can be predicted.

models. It should be noted, however, that in this case the symbolic model resides in the software, and can be used to create both iconic and analogic models on the screen, as shown in Plates 4.1 and 4.2.

Evaluating a design by simulating certain aspects of its performance is one important function of a model, but there are several others. First, models are used to assist in the earliest stages of design, when the problem is being defined and solution concepts generated. Schematics can be used to systematically identify the requirements of a design, and rough sketches enable possible solutions to be quickly explored. Second, models can also be used to communicate information between interested parties, such as designer, client, manufacturer, operator, regulator and so on. In particular the designer needs to inform the manufacturer of the

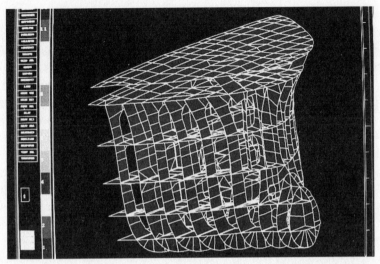

Plate 4.1   A finite element model of the bow of a ship showing the internal structure.
A symbolic model of the structure is stored on disk, and sophisticated software uses this model for detailed structural analysis. The computer can present the model on the screen in a variety of ways, including this iconic visualisation of the internal structure
(courtesy of Mr P. Atkinson, Department of Marine Technology,
University of Newcastle upon Tyne).

Plate 4.2   A finite element model showing stress concentrations.
The same software can present the results as an analogic model, here showing the stress concentrations in one transverse bulkhead of the ship's structure
(courtesy of Mr P. Atkinson, Department of Marine Technology,
University of Newcastle upon Tyne).

required shape of all parts of the product. Traditionally, drawings conveyed this information but, with the advent of CAD–CAM technology, it could equally be communicated via a computer terminal. Where computer-controlled machinery is installed the information is fed directly into the equipment from the design office, without the need for interpretation by any human operator. Lastly, models are also used to store information. Details of a finalised design are kept long after the product has been built and commissioned. These may be needed if a flaw should become apparent once the product has been in service for some time, or they may be used as the basis for another project which can be an adaptation or variant of the first design. Storage of physical models can clutter up a lot of space, and while drawings and diagrams require less room, careful cataloguing is required to ensure easy retrieval of the information when required. This is another advantage of computer models, as the files can be stored on disks, which require negligible storage space and make information retrieval a rapid and simple task.

The use of models in the design process does, however, give rise to one significant problem. Designers unavoidably use several different models on a single design project, as individual models are required to evaluate different aspects of the design. Testing a model will indicate how that aspect of the design may need to be modified, but if the test model is adjusted, all the other models must be similarly adapted. If this is not done the subsequent tests on other models will be carried out on an out-of-date design. This problem is inherent in the use of drawings, as each individual drawing is a model in its own right, and unless there is a rigorous system of management to maintain *version control*, incorrect drawings can be used to obtain data, or even released to the manufacturer and so used on the shop floor. The cost of putting right mistakes due to releasing the incorrect version can be high and may, on occasion, be so great as to jeopardise the viability of the project.

The more sophisticated computer models avoid this problem by storing all the design information on a single database, so that there is in practice only one central model. The necessary information is extracted from this central database to create specific views, or to carry out detailed analysis of particular aspects of the design. In the event that modifications are made, then the central database is updated automatically, ensuring that a consistent version is used at all times. If many designers are able to access the same database simultaneously there is still a problem of concurrency, but this can be addressed by including a product data management system within the software.

The use of models: a case study

Boat building was discussed at the beginning of this chapter as an example of a craft-based industry that manufactured products apparently without ever designing them. Today, both boats and ships are intensively designed, and an examination of the growth of design in this industry reveals that it is intimately bound up with the increasing use of models and the resulting greater sophistication of the simulation process. A brief summary of this development can be used to highlight the link between modelling and design.

The first models used in shipbuilding were 'half-models', physical carvings of one half of the proposed hull shape. A wooden block was whittled away until the builder and customer were satisfied with the hull form, their evaluation being made purely by inspection. Interestingly these models were not used to test performance, but purely to assist in the production of the vessel. The finished half-model was carefully cut into a series of segments which, when separated, revealed the precise shape of the hull at regular intervals along its length. Scaling up these shapes enabled full size patterns to be made, around which the hull could be built. Clearly, experience gained from examining the section shapes revealed by the half-models allowed draughtsmen to draw the proposed vessel instead, and so the two-dimensional models evolved which are still used today, and which are referred to as a set of drawings.

Although both the half-model and the drawings accurately represented the vessel's geometry, in the sixteenth and seventeenth centuries it was not understood how they could be used to evaluate performance, as evidenced by the *Wasa* in Sweden and the *Mary Rose* in England, both of which capsized on their maiden voyages. Only with the development of physical theories and mathematical models in the eighteenth and nineteenth centuries were aspects of a proposed vessel's performance able to be evaluated and its behaviour predicted before construction was begun. Some performance aspects, such as stability, could be calculated from the drawings but others, such as the vessel's resistance, could not. For this reason, in the latter half of the nineteenth century physical models were introduced again; only this time it was to simulate the behaviour of the proposed vessel by carefully observing and monitoring a scale model as it was towed along a tank. Tank tests of this nature are still used today, as are model tests of individual items such as fin keels and propellers, as illustrated in Plates 4.3 and 4.4. It should be noted, however, that the use of scale models to predict performance is problematic and often requires

Plate 4.3   A ship's propeller.
This is a large and expensive item; therefore its performance must be accurately
predicted
(courtesy of the Emerson Cavitation Tunnel, Department of Marine Technology,
University of Newcastle upon Tyne).

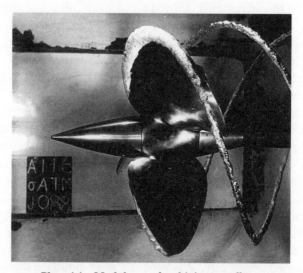

Plate 4.4   Model test of a ship's propeller.
Scale models of propellers can be observed in a test tunnel to ensure that cavitation
will not occur under operating conditions, as this would severely affect the propeller's
performance. In the test shown, cavitation is occurring at the leading edge of each
blade, the resulting voids forming a helical trail behind the propeller
(courtesy of the Emerson Cavitation Tunnel, Department of Marine Technology,
University of Newcastle upon Tyne).

sophisticated calculations to interpret the results. This is because the attributes of a design do not necessarily scale directly with the dimensions. The obvious example to demonstrate this is that of surface areas, which clearly scale with the square of the dimensions. The scaling of other characteristics, such as strength, or energy inputs and outputs, have to be considered with care.

During the twentieth century the theoretical understanding of many phenomena relevant to the design of ships has progressed enormously. The disciplines of materials science, structural analysis, fluid dynamics and thermodynamics have all produced theories from which mathematical models can be developed. Ship designers progressively incorporated these into their procedures, thereby improving the efficiency of their vessel's structure, machinery and propeller, and also improving performance in terms of seakeeping and resistance. In the first half of the century these mathematical models were evaluated by hand calculations, aided by log tables and slide rules. The introduction of computers enabled these calculations to be carried out more rapidly and the sophistication of the mathematical models to be enhanced. In recent decades software has been developed which, for example, enables structures to be evaluated with finite element methods, ship motions to be calculated from computational fluid dynamics packages and survivability to be estimated using flooding simulations.

The development of these analysis packages has progressed in parallel with specialist graphics software which enables the complex geometry of a ship hull to be accurately defined and presented on a computer monitor. When the hull model is combined with the output from other graphic design packages the visualisation of the internal and external appearance of the vessel, from any user-specified point, is possible. In some marine applications virtual reality (three-dimensional imaging) has been used to evaluate specific aspects of the design, such as the arrangement of pipes and equipment in an engine room to ensure adequate access for operation and maintenance. This precise definition of the geometry of the hull and structure is used to increase manufacturing efficiency, as information can be fed to numerically controlled machines which cut out the required plates and structural components automatically (Plate 4.5).

To conclude: shipbuilding, which grew out of the craft-based industry of boat building, is now a highly technological industry which uses computers widely to aid both the design and manufacture of its products. The models used in the design of boats and ships have evolved from simple wooden carvings to the powerful integrated software packages of today,

Plate 4.5   Numerically controlled machine cutting steel plates in a shipyard
(courtesy of Dr R. Kattan, Department of Marine Technology, University of Newcastle upon Tyne).

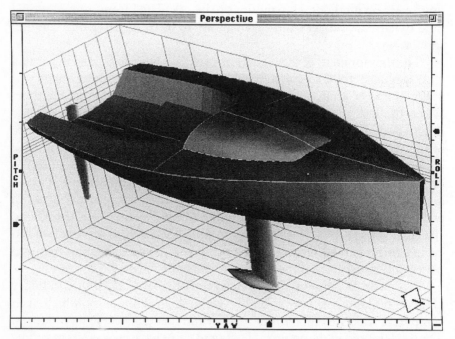

Plate 4.6   A computer visualisation of a proposed yacht design
(courtesy of Christian Stimson Yacht Design).

which allow the designer rapidly to generate a hull form and to evaluate many aspects of its performance. With the same software, accurate and attractive representations of the finished product can be produced to confirm that the purchaser and operator will be satisfied (as shown in Plate 4.6), and then conventional drawings or numerical data can be produced for the manufacturer.

Summary

In the industrial age craft-based manufacturing has evolved into modern production systems by the gradual transfer of decision making from the manufacturer to the designer. In this process the designer's role has expanded such that the design team may now be responsible for the success of all aspects of the product throughout its life cycle. To achieve this effectively the importance of accurately modelling the proposed design has increased, and so a wide variety of domain-specific models have been developed which simulate the behaviour of particular products and enable designers to evaluate their performance. In the information age many of these models are computer-based, and produce vast amounts of data which have to be processed carefully.

By the late 1950s the changing nature of industry, which resulted in the increasing importance of design, led to the setting up of an inquiry which reported in 1960 (the Feilden Report: Feilden, 1963). Among other conclusions, Feilden indicated that in the increasingly complex world there was a need for greater co-operation between the design team, and for more efficient management of the design effort. Since this time the design process has been the subject of extensive academic study, which has produced a considerable body of knowledge, and wide-ranging debate. The progress of this debate in the latter half of the twentieth century is outlined in the next chapter.

Chapter 5
Recent developments in theories of design

Introduction

In the early 1960s, there was broad consensus among academics and industrialists in the UK, the USA and continental Europe that design activity was generally being mismanaged, and that the process of design was starting to become a major bottleneck in the design and manufacturing process. As a consequence of the strength of feeling about the issue, a conscious decision was made to undertake research to identify and develop systematic procedures for the overall management of the design process, as well as systematic techniques to be used within such a process (Jones and Thornley, 1963). The underlying logic of this decision to take action was compelling: without an adequate base of scientific principles, engineering design education and practice would continue to be guided by the empiricism, intuition and specific experience which were the recognised causes of the existing problems (Dixon, 1987). It did not become clear until much later, however, that design was in a pre-science phase. In other words, there was a lack of agreement between researchers and practitioners as to what would constitute a coherent body of knowledge relating to general laws and theories for design, and knowledge of how designers could apply these. It eventually became clear that a period of directed research was essential if design was to become a mature science.

To reach this level of maturity it was also clear that design would need to borrow ideas from other disciplines that had already achieved a fully-fledged scientific base (Coyne *et al.*, 1990). In this respect two distinct approaches offered themselves:

- Using case studies – an approach that was prevalent in such disciplines as early psychology, prior to the establishment of appropriate and acceptable experimental methods.

- Using models – an approach which is less ambitious than that of proposing theories, being content with just the description or explanation and prediction of phenomena.

Early design researchers took the models route and during the subsequent thirty years three generations of models evolved (Cross, 1993). The initial period of research produced a generation of models that were characteristically systematic in nature, were focused on finding an optimal solution to a problem and were intended to be generic and therefore applicable in a wide variety of domains. Further research during the 1970s and 1980s produced two distinctly different types of model, these arising from the disciplines of engineering and architecture. Both the engineering and the architectural models recognised the importance of the specific problem being addressed and the importance of devising a limited range of satisfactory (as opposed to optimal) candidate solutions. Finally, work since the late 1980s has focused on a third generation of models which are hybrid in nature. They combine ideas from the models and concepts previously generated by both engineering and architectural researchers.

Despite the fact that an extensive amount of research has been undertaken over the last thirty or so years, there is still no agreement on a single model which provides a perfect or even satisfactory description of the design process (Bahrami and Dagli, 1993). As was intimated earlier, more research needs to be undertaken before design becomes a mature science. In the meantime, people will undoubtedly continue to propose models of design since these seem to offer a viable means of advancing knowledge. Indeed, although models do not actually constitute a theory, a theory can emerge when there are feasible explanations as to why a model behaves as it does (Dixon, 1987). In other words, it is an acceptable scientific approach to hypothesise a model and then collect and analyse experimental data to ascertain how robust the model is and the extent of its validity. Consider the comments of French (1985), who was Professor of Design at Lancaster University in the UK, on the issue: 'Constructing block diagrams is a fashionable pastime, especially in fields like design where boundaries are imprecise and interactions legion, so that any ten experts will produce ten (or a hundred). They will all be different, and all valid' (p.1).

Hopefully, better and more accurate models will gradually evolve until the point is reached where a theory can emerge. Then, when the models of the day properly reflect the real-life phenomena they are intended to represent, design will be considered a mature science. Popper

(1963) has also noted that the notion of using conjecture (i.e. hypothesis) and refutations (i.e. determining the extent of the validity of a hypothesis) is an approach that has been employed in other disciplines which are considered to have successfully made the transition to a mature science. For the immediate future, therefore, researchers in the field are likely to continue proposing alternative models.

Recently some researchers in engineering have begun to explore the case studies approach which was mentioned earlier as an alternative to that of modelling (Wallace and Hales, 1987; Brereton *et al.*, 1993; Fricke, 1993). Although this type of research has been well established in the architectural field, in engineering it is still an unusual approach, and clearly future work will combine the research results from the models and case studies approach in both disciplines. The discussion which follows examines the major developments which have taken place and the contributions of the key authors to date.

Research in the early years

The focus of design research during the nineteenth century and early decades of the twentieth century was on the formal engineering sciences required for design (Reuleaux and Moll, 1854; Erkens, 1928). By the end of the Second World War, however, the first writings on design itself had been published. These promoted the notion of a step-wise approach to design, particularly stressing the importance of analysis and evaluation. The purpose of this work was to rationalise and improve the design process, an admirable ambition which would not be out of favour with people working in the field today.

Notwithstanding the validity and usefulness of these earlier works, it was not until after the Second World War that design began to emerge as a widely-recognised academic discipline. The years immediately following the war had seen an economic boom in the USA and the UK. Demand for new products outstripped the capacity of industry to supply them, with the result that products could be sold irrespective of the quality of their design. But by the late 1950s the market had changed. Consumers were being offered a choice of products, and they could choose those that were well designed in preference to the rest. In addition, for manufacturing industries in the UK and the USA, competition from the Far East was on the increase. By 1962 the problems were becoming so acute that the

UK government commissioned a report. This, the Feilden report which was mentioned in the previous chapter, concluded that design was of paramount importance in the country's economy, and asked for more effective design management, more attention to customer requirements and more co-operation in design teams (Feilden, 1963). In contrast to the British and American experience, the situation in Germany, which had been desperate in the period immediately following the war, had improved so much that design itself had become the critical bottleneck. This situation was unacceptable since it inhibited further increases in the effectiveness and efficiency of that nation's manufacturing operations. These circumstances eventually led those concerned in each respective country to recognise that there was an urgent need to place design on a more scientific footing.

In the UK and the USA the early 1960s saw the first publications on design methods by some of the most prominent authors who have worked in the field to date. At the time, and possibly somewhat surprisingly in the light of later developments, there was a great deal of unity and common ground between the engineering and architectural communities. Indeed the design methods movement, comprising people from both disciplines, was instigated in this era with the first of a series of conferences on design issues being held in London in 1962. The success of the London conference prompted another in Birmingham (1965), which was subsequently followed by a third and fourth in Portsmouth (1967) and Massachusetts (1969). These conferences continued into the 1970s.

The focus in the UK and the USA at the beginning of the 1960s was on the more aesthetic and creative aspects of design, and was *solution-oriented* in that the procedures suggested conceiving a likely-candidate solution to the problem and then using this to derive more information relating to the specific problem being addressed. In contrast, the early pioneers in Germany were concentrating on functions and working principles as intermediate definable steps between problem specification and design concept. These efforts were focused on the domain of mechanical design, and the approach could be characterised as systematic and *problem-oriented* (this last term indicating that the process of solution generation followed on from the abstraction and reconsideration of the problem, rather than on the analysis of an early initial solution concept). In contrast to the work being undertaken by researchers in the UK and the USA, the work in Germany was being driven by the national standards institution (VDI, 1987), which published the first normalised description

71

of the design process. As time went by, however, the work being done in the UK and the USA began to move toward that being done in Germany, so that by the late 1960s the underlying structure of many of the models from all three countries was based on a systems engineering approach. As a corollary, these models tended to assume that the specification of requirements, albeit important, were merely provided to the designers, and that their objective was only to synthesise an optimal solution to these requirements. This assumption proved to be inappropriate, however, and became the point of much debate and disagreement in later years.

With hindsight, therefore, it is possible to conclude that, although they made useful contributions, many of the works of the day only presented loosely coupled generalised lessons and methods that could be applied during the process of designing. Typically, they advocated the application of systematic, rational and scientific methods. It was later pointed out that the tendency in developing these methods was to take the techniques that happened to be available at the time, and force these onto design without carefully questioning their relevance (Broadbent and Ward 1969). Although techniques such as graph theory, set theory and systems analysis were useful, they should not necessarily have been used as the basis for developing generalised design methods.

Despite a certain degree of concern about the generality of some of these first generation models, concepts and contributions did emerge that proved to be more significant and long lasting. One such early contributor was Hall (1962) who was the first to suggest a two-dimensional view of project development, where the dimensions were defined as follows:

- Vertically – indicating the phases in the life-cycle of the product.

- Horizontally – indicating the problem-solving process that take place at every stage of the vertical dimension.

Using these dimensions as the basis for his proposal, Hall's model of the systems engineering process is represented in Figure 5.1.

Subsequently, Asimow (1962) developed Hall's ideas to derive his model of the design process. Asimow was already a well-respected academic, being the Professor of Engineering at the University of California (Los Angeles) at the time. He derived a similar two-dimensional model, which is still referred to today, specifically for design. Asimow defined the vertical axis of his model as the *morphological* dimension and the horizontal axis as the *problem-solving* dimension.

Phases of the coarse structure (time) / Steps of the fine structure (logic)	1. Problem definition	2. Value system design (and develop objectives and criteria)	3. Systems synthesis (collect and invent alternatives)	4. Systems analysis (deduce consequences of alternatives)	5. Optimisation of alternatives (iteration of steps 1–4)	6. Decision making (application of value system)	7. Planning for action (to implement next phase)
1. Problem planning							
2. Project planning (and preliminary design)							
3. System development (implement project plan)							
4. Production (or construction)							
5. Distribution (and phase in)							
6. Operations (or consumption)							
7. Retirement (and phase out)							

Figure 5.1 Hall's model of the systems engineering process.
This has two dimensions, the vertical dimension corresponding to the origination stages in the life cycle of a product, while the horizontal dimension relates to the problem-solving activities that take place in every stage of the vertical structure.
(Reprinted from A. D. Hall, *A Methodology for Systems Engineering*, 1962, © Van Nostrand.)

According to Asimow, morphology is intended to relate to the chronological structure of a design project (i.e. the primary phases of the design process), while problem-solving is intended to relate to the highly iterative and cyclical activities associated with generating candidate solutions.

Asimow then went on to propose a useful framework wherein such models of the design process could be placed. At the highest level of this framework he suggested that there is a philosophy of design which encapsulates the fundamental principles and concepts which are relevant to whole classes of problem. According to Asimow, the existence of a philosophy leads to the development of theories, laws and rules, and to methods of applying these, which can be considered to represent the discipline of design. Distinct from philosophy, the discipline of design is the second level and deals with recognisable categories of problem. This represents the level in the framework where models can actually be located. The discipline therefore represents an intermediate intellectual structure which can help guide the attack on categories of problems. Finally, at the lowest level of the framework are particular strategies or tactics which are derived from the discipline. These tactics represent the particular approach that a design practitioner would evolve, through experience, for tackling specific types of problem.

Most importantly, Asimow also suggested that because philosophy encompasses an individual's beliefs and values, it is coloured and biased by their experiences and idiosyncracies, and therefore there cannot be just one philosophy of design. As a corollary to this, there is evidently not just a single discipline, or tactic either, and this explains why there is such diversity of opinion in the respective research communities as to the form of a single model of design.

As suggested above, further research and debate will be required before broad agreement is reached as to what constitutes the philosophy, disciplines and tactics of design. As Kuhn (1970) has indicated, however, design will not be considered a mature science until there exists this coherent tradition of scientific research and practice which embodies law, theory and application, these terms corresponding to Asimow's philosophy, disciplines and tactics.

The birth of design methodology

During the late 1960s and early 1970s some of the early influential figures in the field began to reject the whole concept of a science of design, due to the lack of success in applying systematic methods to design practice. In particular, Alexander (1971) was very critical, even of the new name that had been given to the continuing research:

> I never even read the literature any more . . . I would say forget it, forget the whole thing . . . If you call it 'It's a good idea to do', I like it very much; if you call it 'A method', I like it but I'm beginning to get turned off; if you call it 'A methodology', I just do not want to talk about it.

Despite the fading support of some researchers, particularly from the architectural community, there was still sufficient interest from other quarters for significant developments to be made. The first of these was a hypothesis by Simon (1969), who, in the Karl Taylor Compton memorial lecture at the Massachusetts Institute of Technology, stated that the goal of design activity was not necessarily to find a single optimal solution. The goal, Simon suggested, was to generate a number of candidate solutions which might not be optimal but which individually satisfied all requirements to a greater or lesser extent. Indeed, Simon actually coined the term *satisficing* to describe such candidate solutions.

Further, and perhaps more significantly, Rittel and Webber (1973) hypothesised that there were two types of design problem and classified these as *ill-structured* and *well-structured*. Put in simple terms, Rittel and Webber suggested that only well-structured problems could be described exclusively in terms of numerical variables such that available algorithms permit the optimum solution to be found. Rittel and Webber went on to conclude that many design problems, if not all, did not fit this description and so should be called ill-structured. The pernicious character of such problems implied that the application of the analytical approach, typified by the early design models, was therefore very limited.

These new notions contradicted the existing shared model of design and an extremely turbulent period of debate followed. Deep divisions began to emerge between the architectural and engineering communities in the UK and the USA regarding the nature of problems and the form that a model of design should take. This schism caused research activity to branch off into two distinctly different, if still related, directions. Interestingly, the architects followed a path which can be loosely related to the problem-solving dimension of Hall's and Asimow's earlier models. In contrast, engineering research continued along the route which had begun to evolve at the end of the 1960s. This particular route can be related to the morphological dimension of Hall's and Asimow's models and, as mentioned earlier, was a similar approach to that being taken by the German researchers.

The new developments in the 1970s and 1980s were in the field of architecture, the first significant contribution being made by Hillier *et al.* (1972), who began to question the whole notion that designers should resist bringing their own preconceptions to bear on a problem. Broadbent and Ward (1969) similarly questioned the idea that a designer would not use personal experiences to influence the nature of a solution offered to solve a particular problem. As a result of his disillusion, Hillier proposed the novel idea of a conjecture–analysis model which reflected the fact that a designer would actually prestructure problems in order to solve them, that is, they would use previous experiences and knowledge to influence the nature of the solution. Darke (1978) extended this concept using research findings from the field of psychology to promote the generator–conjecture–analysis model. This model identified the predominant role of the *primary generator* in making the creative leap from design problem to potential solution. Darke coined the term 'primary generator' to describe a simplified problem definition which she had found designers used in the early stages of design in order to stimulate the conjecture of possible solutions. A symbolic representation of Darke's model is shown in Figure 5.2. The view represented by this model reflected Darke's observations of actual architectural design practice, where a designer uses potential solutions at the early stages of the process in order to elicit more information about the problem from a client.

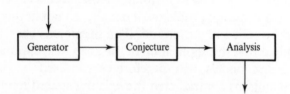

Figure 5.2   Darke's partial map of the design process.

The basis of this model is that the designer, when faced with a complex problem, does not start with a full and explicit list of the factors to be considered. Instead the variety of potential solutions is reduced to a smaller class that is cognitively manageable. To do this, the designer fixes on a particular objective or small group of objectives, usually strongly valued and self-imposed, for reasons that rest on subjective judgement rather than being reached by a process of logic. These major aims are known as the primary generators and are what give rise to a proposed solution or conjecture. This makes it possible to clarify the detailed requirements, as the conjecture is tested through analysis to see how well the aims are met.

Plainly, the architectural view was in direct conflict with that of the engineering fraternity, where emphasis was generally placed on the need for an exhaustive specification of requirements prior to initiating the design process. The work on the systematic methods of the 1960s, particularly the later work influenced by German research, had led to the evolution of a model which was principally focused on Asimow's morphological dimension. This model, presented in its simplest form by Lawson (1990), comprised three stages, these being analysis, synthesis and evaluation, as shown in Figure 5.3.

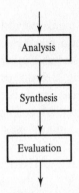

Figure 5.3 Lawson's three-stage model of the design process.
This simple model of the design process comprises the three stages of analysis, synthesis and evaluation. According to Lawson, analysis essentially involves the ordering and structuring of the problem, synthesis the generation of solutions, and evaluation the critical appraisal of potential solutions against the objectives identified in the analysis stage.

The model is broadly similar to Darke's (Figure 5.2), but it is important to recognise that in each case the emphasis is distinctly different. The engineering model assumes that the designer is able to construct a specification which is *solution-neutral*, that is, that the specification can be sufficiently abstract to allow designers to exploit their creative talent fully without necessarily prejudicing the end result with existing knowledge of probable solutions. In contrast, the architectural model promotes the idea that it is just this existing knowledge which will provide the creative spark and allow the designer to make the leap from problem to candidate solution. The architectural model enables the designer to elicit gradually more and more information from the client, and so learn more about the client's real requirements by proposing conjectured solutions and listening to the response. From this analysis of these contrasting models, it is possible to relate them directly to the two dimensions of Asimow's model, that is, the engineering model is related to the morphological dimension and the architecture model to the problem-solving dimension.

Another significant difference between the two models is the level of feedback and iteration that is presented. The prominence of conjecture in the architectural model (generator–conjecture–analysis) inevitably implies a more cyclical procedure than that suggested by the engineering counterpart (analysis–synthesis–evaluation). Indeed all the models developed in the architectural field since the early 1970s have exhibited this cyclical characteristic, which is not so prominent in the engineering models developed during the same period. One particularly interesting example of such an architectural model was proposed by March (1976), at around the same time as Hillier was promoting his ideas. March argued that the essential logic of the design process required that solution concepts were produced not only by inductive and deductive reasoning, but also by productive reasoning. March's model is shown in Figure 5.4.

The basis of March's argument was that the two conventionally understood forms of reasoning, inductive and deductive, only applied to the analytical and evaluative activities of design. March found it

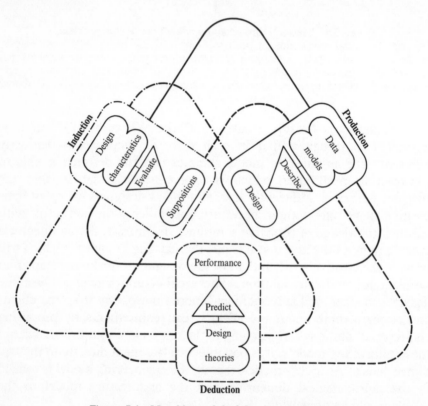

Figure 5.4   March's model of the design process.
March's radical model of the design process recognises the solution-focused nature of design thinking. In this model, the phase of *productive* reasoning draws on a preliminary (and possibly vague) statement of requirements, and some presuppositions about solution types, in order to conceive a potential design proposal. From this proposal it is possible to analyse *deductively* the performance of the candidate solution. Finally, from the predicted performance characteristics of the design it is possible to evaluate *inductively* further alternatives, possibly leading to changes in the design and changes or refinements in the design proposal itself.

(Reprinted from L. J. March, *The Architecture of Form*, 1976, © Cambridge University Press.)

impossible to envisage how these types of reasoning could be associated with the most important activity in design, that of synthesis. He went on to conclude that the kind of thinking which resulted in design proposals should be termed 'productive' reasoning. From this he derived his model in which, although there are no arrows actually shown, the general direction of the process is clockwise. In this methodology the designer, at the first stage of productive reasoning, draws on the (possibly vague) problem specification and uses existing knowledge to conceive a candidate solution. On the basis of this initial proposal, the designer can deductively analyse the design and then inductively evaluate its performance. In collaboration with the client the problem specification may then be further refined and/or evolved. This model, as one can see, is highly cyclical in nature.

Meanwhile, as Hillier, March and others evolved their ideas towards what has been referred to as the *type* model, engineering research progressed on the basis of the more prescriptive foundations laid during the 1960s. Pahl and Beitz (1984), who were eminent German academics with a long association with the design research community, were early contributors to the field. Their work, which was based on the earlier German standards (Erkens, 1928), had its roots in the design of mechanical systems. As a consequence the general approach advocated by Pahl and Beitz was strongly systematic. The topmost level of their very detailed model is reproduced in Figure 5.5 and conveys the general impression of the approach.

This original contribution was followed by other similar models (Hubka, 1982), all of which promoted the analysis–synthesis–evaluation approach. Later, during the 1980s, other people began to extend these ideas to cover more of the product life cycle. A British Government report (Fielden, 1963) had concluded that more attention should be paid to managing the entire design process, and that there should be more co-operation between the designers in a team. These new models reflected some of these concerns, as they indicated the idea of teamwork by showing parallel activities, as in concurrent engineering, and extended the morphological scope to cover more of the product life cycle (Andreasen, 1991; Ullman, 1992). They also incorporated outside influences which impinge on design activity (Andreasen, 1991).

One proposal of particular interest was produced as part of a wide-ranging initiative and had, as its focus, the goal of improving design teaching. The SEED (Shared Experiences in Engineering Design) programme was initiated by an association of design lecturers in Britain,

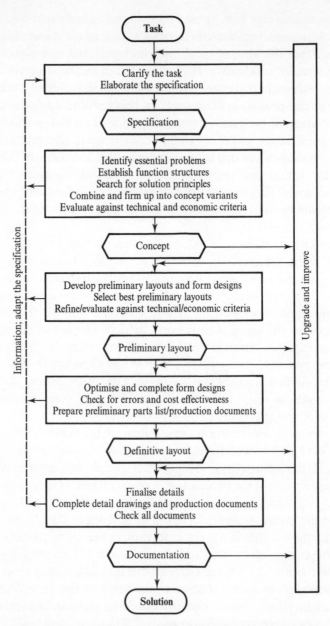

Figure 5.5   The Pahl and Beitz model of the design process.
This shows the process comprising a number of steps wherein the main phases include clarification of the task, conceptual design, embodiment design and detail design. At every step a decision must be made as to whether the next step can be taken or whether previous steps need to be repeated.

(Reprinted from G. Pahl and W. Beitz, *Engineering Design*, 1984, © The Design Council.)

and has emerged as a significant force for the improvement of design education. The aim of this initiative was to improve the quality of design education by providing a high profile forum for the discussion of matters of concern. The engineering model adopted by SEED as the basis for a common teaching tool was that put forward by Pugh (1990), head of the Design Division of the University of Strathclyde and a prominent figure in the development of the SEED programme. Pugh's model is reproduced in Figure 5.6.

Thus, despite a common origin, toward the end of the 1980s significant differences existed between the engineering and architectural communities. These differences were manifested in two distinct types of model, whose individual characteristics are briefly summarised as follows (Cross and Roozenburg, 1992).

Engineering-type models	**Architectural-type models**
Linear, sequential process	Spiral, cyclical process
Sequence of stages	Cycle of cognitive processes
Prescriptive	Descriptive
Exhaustive evaluation of requirements	Use of existing knowledge
Problem structure is tree-like	Problem structure as lattice-like
Assumes well-defined problem	Assumes ill-defined problem

The emergence of the hybrid model

Although the research effort has now extended for more than thirty years, it could be argued that the field of design methodology is still in its infancy, and that further work will be required before a consensus is reached. However, in the 1990s there is the emergence of a third phase in the evolutionary process, with the introduction of hybrid models. Such models represent a recombining of the engineering and architectural schools of thought, and derive their basis from the solid contributions of earlier authors such as Asimow and Simon. The explanation for the emergence of this third generation of models can be found in the criticisms that have been levelled at both schools. Some of the more disparaging comments that have been made regarding the engineering models follow (Cross and Roozenburg, 1992).

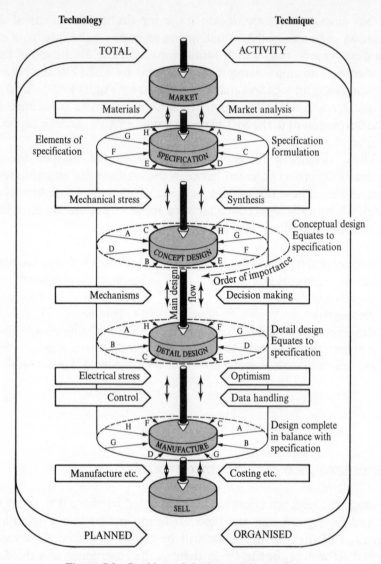

Figure 5.6   Pugh's total design activity model.
In this model the design process is located within the framework of planning and organisation. As can be seen, the model comprises a broad design core which is always enveloped by product design specification. Moreover, the model shows two types of input to the design core: discipline/technology dependent and independent.
(Reprinted from S. Pugh, *Total Design*, 1990, © Addison-Wesley Publishing Co. Inc.)

- They have been developed primarily for the design of technically innovative systems (i.e. paying too much attention to the concept phase of the design process and ignoring the fact that many projects commence from known, proven concepts).

- In practice the behaviour of engineering designers seldom resembles the behaviour prescribed in the model. It should be noted, however, that this does not disqualify the models because they can be considered a prescriptive approach that is intended to structure, and not to predict, design behaviour.

- The partitioning of the design process into a large number of small steps might lead to an uncontrollable explosion of possible solutions, and the function structures that result may be of limited practical value.

Plainly, there has been insufficient time for the research into hybrid models to evolve far, but one significant contribution has already been made. Cross (1989), who worked in the Design Discipline in the Faculty of Technology at the Open University in the UK, produced a model which goes some way to integrating the more useful ideas that lie behind both the engineering and architectural models. This model is reproduced in Figure 5.7.

At first sight this is another model that appears confusing, but examination reveals that it has both prescriptive and descriptive traits, and that it is not too difficult to disentangle and understand. First, it assumes that a symmetrical relationship exists between problem and solution, and between sub-problems and sub-solutions. The purpose of including this relationship is to demonstrate that problem definition is often dependent upon solution concepts. Second, it assumes that a hierarchical relationship exists between problem and sub-problems, and between solution and sub-solutions. The reason for this is to demonstrate that an important aspect of problem clarification is decomposition into sub-problems. Finally, within the overall framework, the model incorporates a set of design activities and methods which are intended to promote and assist in the design process.

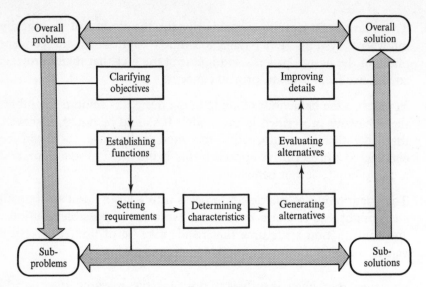

Figure 5.7   Cross's model of the design process.
Cross's model comprises seven stages purposefully positioned within the symmetrical problem/solution model. The model integrates the procedural aspects of design with the structural aspects of design problems. The procedural aspects are represented by the sequence of methods while the structural aspects are represented by the larger arrows showing the commutative relationships between problem and solution and the hierarchical relationships between problem/sub-problems and between sub-solutions/solution.
(Reprinted from N. Cross, *Engineering Design Methods*, 1989,
© John Wiley & Sons Ltd.)

Summary

The academic field of design has evolved rapidly over the past thirty years, and in the most recent generation of hybrid models continues to produce further insights. Although a universally accepted integrated model is still some way off, people are offering proposals which represent attempts to move towards one (Tovey, 1986). At present, however, the field still suffers from a lack of support by design practitioners, as the models offered to date have had little acknowledged practical application (Cross and Roozenberg, 1992). Indeed, the one criticism that has to be levelled at all design models is that there is not the widespread acceptance in industry that is essential to give credence to the theoretical studies. This is in part due to the historical fact that many industry-based engineers received little, if any, formal training in design studies. Indeed, the formal university education of many engineers still lacks a grounding in the fundamentals of design. This is a situation that must be rectified if widespread acceptance is eventually to be achieved. In the UK, however, since the late 1980s the Government has sponsored a number of initiatives to promote design research and practice. These have ranged from minor initiatives such as the publication of a series of best practice brochures, to major capital-intensive programmes to support, for example, Business Links organisations and Engineering Design Centres. All these initiatives have been aimed at shifting people's focus to recognise that design is an extremely important factor in determining a business's success.

Only time will tell if the latest developments will result in a model which is supported by academics and practising industrialists alike. In the meantime, though, as stated at the beginning of this chapter, it seems that the most viable way to continue to make advances in the field lies in developing the hybrid models of design. Furthermore, the research currently being undertaken using a case studies approach may reveal additional insights which could be incorporated into a future generation of models, and so move our understanding onto a more mature level.

Section III
The Practice of Engineering Design

Chapter 6
Practical issues in design

Introduction

The preceding sections of the book have addressed various issues relating to the environment of engineering design, including those elements which have a bearing on how design is practised, and have also reflected upon some of the theoretical propositions which have been made in an attempt to promote improvements in the understanding and practice of engineering design. In this chapter the relationships between design theory and design practice will be explored more fully, including how the latter is conditioned by the environment of design. In particular, the hierarchical nature of design problems, specifications and decision making will be discussed.

Design theory and design practice

In Chapter 5 we discussed the salient features of the engineering design models that have been proposed by prominent contributors to the field in recent decades. Throughout the progressive development of these models there has been a convergence on the use of certain basic features (Pugh, 1986a; Cross and Roozenberg, 1992). These include the breakdown of the design process into conceptually distinct stages or activities, leading to certain intermediate results (specification, functional structure, working principle, preliminary layout, detail documentation, and so on) and the implied notion of linearity, albeit with varying emphasis on iteration and interaction.

It is the case, however, that there has not been the widespread acceptance of these models in industry that might have been anticipated. Indeed, the nature of design activity seldom resembles that prescribed by the engineering design models. Hales (1987) has therefore concluded that

> despite a long history of innovative engineering design in industry and development of many prescriptive methods and models, the engineering design process is not yet considered well understood or adequately exploited . . . there is a mismatch between the design process as it is currently modelled in theory and what actually happens in practice (p.12).

Practitioners in industry most typically tend to develop their own procedures to deal with design projects. These usually tend to be expressed in terms of the overall approach to be adopted on projects and deal with the broad organisational and managerial concerns rather than the more specific issues of design methodology. In other words, procedures usually address what is required to be done as distinct from how it should be done. A company's procedures may be the result of conscious attempts at developing improved practices, or inherited through years of experience of what have been found to be the most successful approaches. They may also be expressed in various ways and presented in differing degrees of detail. For example, in some companies, procedures are formally written down (often being incorporated into an approved quality assurance manual), whereas in others they are not.

Several reasons may be assigned to the apparently limited use of the engineering design models in industry. First, there is the question of awareness. Experience from research activities by some of the authors indicates that few practitioners have more than a basic awareness of the engineering design literature. However, while this may be viewed as a failure on the part of engineering design practitioners to seek out the relevant literature, it must also be seen as a reflection of the limited importance which historically has been accorded to the formal teaching of design methodology in engineering courses

This limited application of the engineering models may also be viewed as a deliberate rejection based on their perceived lack of relevance and other specific methodological criticisms. While all design methodologists recognise the differing nature of design problems, and acknowledge three distinct types of design activity (defined in Chapter 2 as original, adaptive and variant design), most of the prescriptive models have been developed with a view to resolving original design problems, and this despite it being known that much design activity is of a non-original nature. In defence of this, several reasons are put forward by the models' proponents: original design problems are seen to represent the most pervasive and broad type of design in terms of the variety of possible activities; non-original design is argued to be a subset of original design, the overall process remaining the same, with only a change in emphasis being required; non-original design may also involve elements of original design in some parts of the product being redesigned.

A second criticism of certain models is that in prescribing improved approaches, erroneous assumptions of ideal or conducive circumstances are made. The main emphasis with these models is typically on the quality of the design process and solution, and their implicit assumptions that variables such as time, available resources, quality of information, quality of the designers, management and working conditions are not normally constraining factors.

Finally, contrary to the assumption that the engineering design process proceeds from an abstract analysis and problem formulation to the articulation of solution concepts, engineering design in practice is more often an interactive, recursive process which frequently relies on the use of existing design concepts to anticipate possible solutions in which conjecture and problem specification proceed side-by-side rather than in sequence. This difference of approach is at least in part attributable to the constraints which are usually placed upon the designer by those decision makers empowered with the responsibility for establishing the project's framework. As discussed in Chapter 2, in most instances it is in a company's interest to place some constraints on the designer. Typically these will arise out of concerns for likely sources of risk, and the necessary contingencies required to manage it, and also the implications arising from the type of innovation strategy pursued by the company. It must also be recognised that design projects take place within the constraints imposed by the status of technologies and the availability of resources. However, such considerations are themselves conditioned by broader or higher level influences. Consequently, as discussed in Chapter 1, consideration must be given to factors external to the company, such as those associated with its markets and the wider social and political environment. These are highly likely to exert some influence on both the strategic design and development plans and on the scope of individual projects. Legal requirements, design rules and codes are examples of such external influences.

Design strategy

The key feature which distinguishes between the engineering design models and the approaches commonly adopted by design practitioners is the sequence in which the design stages and activities are planned or executed. Taken collectively these stages and activities represent a *design strategy*. A *stage* can be defined as a subdivision of the design process based on the state of the product under development. A design *activity* is a subdivision of the design process that is based on the procedures. The stages of the design process are therefore completed as a linear sequence, while the activities usually recur regularly, in part or in full, during the design process. A *method* is a systematic way in which one or more activities are performed and, often, the practical means for executing a method is referred to as a *technique*. Activities and methods will be dealt with in Chapters 7 and 8 respectively. However, it is worth noting here that the activities all involve information processing and are usually referred to using terms such as collecting, generating, elaborating, evaluating, modifying and documenting. Similarly, according to Pugh (1990), there are both generic techniques applicable to any application domain (such as analysis, synthesis, decision making and modelling) and there are also discipline-dependent techniques derived for specific areas of knowledge (such as structures, thermodynamics and electronics).

Thus certain differences between the engineering models can be assigned to the way in which they discern divisions of the stages and activities (i.e. the design strategy implied). Some of the models limit the distinction to the main stages, while others extend this to include intermediate steps. Furthermore, some of the models incorporate activities explicitly in combination with the stage division to describe how to proceed during a specific stage. The most fundamental feature of the models in terms of their individual design strategies concerns their methodologies for the translation from problem statement into product definition – this being the core process in design and development projects. The engineering design models, which on the whole are *problem-focused*, concentrate initially on analysis of the problem (i.e. characterised by abstraction steps), followed by a systematic concretisation process. However, the available evidence indicates that designers in practice will often adopt a *product-focused* approach. Such an approach emphasises analysis of the product concept (characterised by the use of solution conjectures to generate a solution concept and so to gain further insights and an improved definition of the problem) and promotes the notion of the solution and problem specification being developed simultaneously. This process is characteristically heuristic in nature, drawing on previous experience, general guidelines and rules of thumb. This is then followed by further analysis and evaluation steps to refine and develop the solution. This basic distinction between the two approaches may be regarded as representing the polar extremes of a design strategy spectrum. Clearly, therefore, contrary to the prescriptive assumptions of most engineering design models discussed earlier (in which design process proceeds from an abstract analysis and problem formulation), design in practice is more typically an interactive, recursive process which relies on the use of pre-structures and the anticipation of possible solutions in which conjecture and problem specification proceed side-by-side rather than in sequence. A minority of the engineering models (e.g. Asimow (1962) and French (1985)) propose product-focused design strategies, as do most of the architectural models. It is interesting to note, however, that these have been developed predominantly on the basis of descriptive studies, as was discussed in the previous chapter.

In addition to the fundamental distinction between problem-focused and product-focused approaches, design strategy may also be related to the kinds of thinking processes involved. Based on the work of psychologists, Cross (1989) has noted three dichotomies of thinking style which are applicable to engineering design. These are convergent and divergent thinking, serialistic and holistic thinking, and linear and lateral thinking. While all these kinds of thinking are necessary and relevant to good design, it is the distinction between convergent and divergent which is of particular importance here. Although the designer's overall aim is ultimately to converge on a detailed proposal, it is often the case that, in realising this, periods of divergence may be required. Divergent design activity is usually most prominent in the early design stages when attempting to generate potential design solution concepts. In addition to the initial generation of potential solutions, an outcome of the evaluation process, which seeks to identify the most promising solutions, can often be to stimulate new ideas or possibilities to improve existing proposals. This cycle between convergent and divergent activities has been referred to as controlled convergence (Pugh, 1990).

In the previous chapter Asimow's hierarchical framework of design was described. This framework views design methodology at three levels: philosophies, disciplines and tactics of design. It will be evident that in prescribing an overall design strategy in the form of a general combination of stages, activities, and methods, most of the models are essentially dealing with the discipline of design. It follows that in prescribing a strategy that is applicable to a whole group of design problems, it is assumed that the designer is able to elaborate a specific approach (or tactic) to resolve a particular design problem. The designer's first task at the outset of any project is to develop a general plan or strategy for the design process (Cross, 1989). However, in attempting to prescribe improved ways of working, some of the engineering design models are deliberately constrained to a particular strategy, albeit a general one, which limits the degree of flexibility open to the designer. Given the wide variety of design problems and constraints which a designer may encounter, a more flexible strategic approach to designing is required which identifies and allows the appropriate kinds of thinking at each stage and for the particular context of a design project. Indeed, to a certain extent a positive feature of some of the more recent engineering design texts (Cross, 1989; Pugh, 1990; Ullman, 1992) is their more pragmatic or balanced treatment of design and the flexibility of approach they accommodate.

A design strategy, when described as both a general approach and the sequence of activities to adopt on a particular project, implies a predefined or explicit plan. Indeed, when dealing with non-original and familiar types of design problem, the design strategy is also likely to be one which is predominantly convergent in nature, and for which the plan is known in some detail. Design, however, will often commence with only a broad idea as to how to proceed. In these instances the strategy adopted, at least in detailed terms, may only be apparent in a retrospective sense. This is most common when dealing with an original design problem, for in this case it may be appropriate initially to adopt a more exploratory approach in order to widen the search for solutions. This could involve an abstractive approach, as proposed by most engineering models or, as the architectural models reflect, existing concepts and solutions could be used to acquire a better understanding of the problem and to promote further ideas. In these instances more divergent types of design activities, methods and thought processes are to be expected, on the basis of which a more explicit approach may be elaborated.

The hierarchical nature of design projects

The decision-making processes and activities associated with design projects can be considered from a hierarchical perspective. At one level, issues relating to the breakdown of the design project, or problem, may be considered. In general, other than for the most simple products, the complex nature of many design problems frequently means that it is simply not feasible to tackle the whole problem at once. In order to make the problem more manageable, and to enable teamwork where relevant, it is often necessary to decompose the overall problem into smaller sub-problems. In practice, and in overall product terms, this is predominantly related to the analysis of a product's functions and systems, and hence the subdivision into its systems, sub-systems and components. Ideally, this should be done in such a way as to make the design problem easier to solve.

Newell (1969) categorised problems as being either well-structured or ill-structured in nature, and Simon (1984) identified their characteristic properties. Expressed simply, well-structured problems have clear goals, often a single correct or optimal answer, and rules or known ways of proceeding to the solution. Such problems, according to Simon (1969), are fully-decomposable or nearly-decomposable into branches of sub-problems and sub-sub-problems, for which there is little interaction between the sub-problems and sub-sub-problems of the respective branches. It is possible therefore to solve each sub-problem independently and then to integrate these to arrive at a solution to the overall problem (Levin, 1984).

A good demonstration of a well-structured problem is the jigsaw, which is usually solved by breaking down the overall problem into a number of sub-problems. These typically include the edge of the picture and its principal areas or visual themes. These may in turn be subdivided further according to the difficulty of the jigsaw. In principle, this is repeated until each piece of the jigsaw represents a detailed problem. The jigsaw is progressively solved by combining the pieces to form the components of the picture, combining these to complete areas of the picture, and hence to complete the picture itself. This is done in the knowledge that in resolving the different parts of the picture (sub-problems), these will always combine to form the total picture (overall problem).

In contrast to well-structured problems, many design problems are recognised as being ill-structured (or ill-defined) in nature. Cross (1989) summarised ill-structured design problems as having the following features:

- There is no definitive formulation of the problem.
- Any problem formulation may embody inconsistencies.

- Formulations of the problem are solution dependent.

- Proposing solutions is a route to understanding the problem.

- There is no definitive solution to the problem.

Ill-structured problems therefore characteristically exhibit interdependencies between sub-solutions to sub-problems, so that a sub-solution which resolves a particular sub-problem may create irreconcilable conflicts with solutions to other sub-problems. This pernicious feature means that it is necessary to iterate selectively around the problem hierarchy to resolve difficulties and validate statements made at the various levels, using solution conjectures, in order eventually to realise an overall solution.

Simon (1984) has argued that the distinction between ill-structured and well-structured problems is not a clear one. With a thorough analysis and problem formulation, some ill-structured problems may be formulated as well-structured problems. Moreover, some sub-problems, and more particularly those relating to specific design tasks, may in themselves be well-structured due to the activities and methods of analysis used to resolve them. However, it is worth noting that there is no guarantee that optimal sub-solutions will combine into an overall optimal solution (Luckman, 1984).

We have already identified that designers in practice will often adopt a product-focused approach. Moreover, Cross (1989) has suggested that these product-focused strategies are perhaps the best way of resolving design problems: 'In most design situations, it is not possible or relevant to attempt to analyse "the problem" *ab initio* and in abstract isolation from solution concepts' (p. 30). Cross argues that although there may be a logical progression from problem to sub-problem and from sub-solution to solution, there is a symmetrical, commutative relationship between problem and solution and between sub-problems and sub-solutions, involving iteration between problem and solution, as illustrated in Figure 5.7 in the previous chapter.

As mentioned above, the notion of decomposing a design problem into sub-problems is consistent with the subdivision of the product into its system, sub-system and component levels as applicable. The design activity at each of these levels will typically vary in originality. Indeed, Pugh (1990) suggests that while many overall concepts are relatively static, there are often opportunities for innovation at the sub-system and component levels. This comment leads to two separate observations.

First, such opportunities to innovate are potentially a significant source of work satisfaction for the designers. Second, different design strategies may be required within the problem hierarchy to solve problems with different characteristics. It therefore follows that the subsets of design projects must be managed carefully, as aspects of both the problem-focused and product-focused approaches may be required in a single design situation.

So far the discussion has concentrated on the fact that many of the engineering design models focus on a design process that starts from an initial problem formulation and ends with the completion of the detailed product definition. In these models all influencing factors are assumed to have been included in the given problem brief. However, design activity does not take place in isolation, but is influenced by a number of individuals and organisations within a decision-making hierarchy. This is reflected in a few engineering design models which embrace activities relating to the marketing and product planning processes. However, an even broader perspective on design projects can be given by examining the extended decision-making hierarchy, which includes management of the overall project, project initiation and product strategy. With this view it is possible to observe the changing responsibilities for decision making, as they move between senior management, those responsible for project management and the individuals assigned to the project, including the designers.

At the highest level are issues of product development and marketing strategy, and these are the primary concern of senior management. Senior management also has concerns prior to, and during, individual projects with regard to their approval and review. Below this are the implementation level decisions relating to project initiation or tendering activities, and the actual execution of the project itself. These decisions are of two types. First, there are the project planning and control decisions, the responsibility for which will depend on contextual factors relating to the firm and project. Second, there are the routine operational decisions, made by the project manager and individuals assigned to the project, including the decision-making processes of individual designers.

As discussed in Section I of this text, designers' decisions are clearly bounded by the form of the design requirements and any constraints specified, but they are also related to the hierarchical nature of design problems and the design strategy adopted to solve different problems within the hierarchy. It is reasonable to suppose that at the overall product level, decisions made regarding design strategy will be most strongly influenced by the most senior designers connected to the project. At progressively lower levels, which concern problems of a more detailed nature (i.e. sub-systems and components), the type of design approach taken will be determined increasingly by the vagaries of individual designers. For this reason designers should ideally be skilled in all the types of design thinking which were mentioned earlier in relation to design strategies. However, as individuals, designers will have specific aptitudes and therefore preferences as to which design approach they should adopt (in general engineers are most proficient at the convergent, serialist and linear forms of thinking). It would be inappropriate for senior designers and managers to formulate the design approach without consideration of the individual designers. Moreover, in team situations, the selection of designers to suit the type of project and the stage of the project may be an important consideration. This is due not least to the fact that individuals' characters have an impact on the dynamics and performance of the team and therefore its chances of successfully concluding a project.

The relationship between the project planning hierarchy and the design strategy hierarchy is illustrated in Figure 6.1. Due to their differing perspectives, they are in principle distinguishable from each other at the overall problem and project levels. Only at the levels of detailed activities and methods will they be one and the same. This does not imply that they are disassociated from each other. On the contrary, choices relating to one are likely to impact on the other.

Project and product design specification

An important feature of the decision-making hierarchy concerns the activities relating to the initiation of projects. In Chapter 3 it was indicated that projects start in many different ways depending on the circumstances. Often a design project stems from an idea, identified need, or problem. This may result in an internal development project. Alternatively, if the instigator requires to subcontract part, or all, of the design of the product, then competitive bids will be solicited with a view to placing an order. Although there are similarities, the distinction between these two basic types of project is significant. Of particular importance are the means for specifying both the project and the product design requirements.

A product development project is often initiated in response to an idea

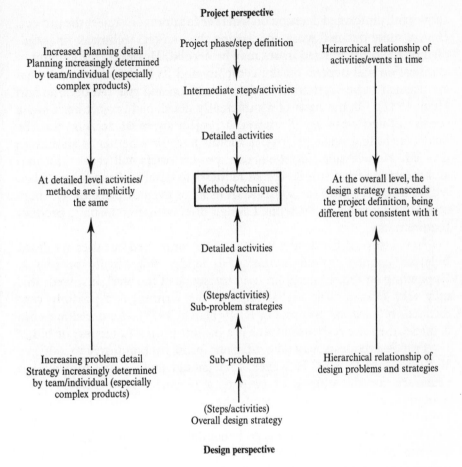

Project perspective

Increased planning detail
Planning increasingly determined
by team/individual (especially
complex products)

Project phase/step definition

Heirarchical relationship of
activities/events in time

Intermediate steps/activities

Detailed activities

At detailed level activities/
methods are implicitly
the same

Methods/techniques

At the overall level, the
design strategy transcends
the project definition, being
different but consistent with it

Detailed activities

(Steps/activities)
Sub-problem strategies

Increasing problem detail
Strategy increasingly determined
by team/individual (especially
complex products)

Sub-problems

Hierarchical relationship of
design problems and strategies

(Steps/activities)
Overall design strategy

Design perspective

Figure 6.1   Relationship between project planning and design strategy hierarchies.

or request. The initial product idea will usually include a brief description of a proposed new product, outlining why it might have potential for the organisation. A product proposal (or project brief) expands this, confirms its potential, and aims to define the work to be carried out. Therefore it should briefly outline the project's objectives and the intended market for the product, as well as providing preliminary estimates for project costs, capital requirements, and financial implications in terms of indicators such as turnover and profitability.

Upon approval of the product proposal a feasibility study or preparatory study usually needs to be undertaken. This is necessary if sufficient information is to be made available to senior management to allow a fully informed decision on whether to approve or reject the project. Hence, more detailed work concerning the market, technical, financial, planning and risk-related issues may be needed. However, the amount of clarification will depend on the company and its background, and what are deemed to be acceptable levels of ambition and risk (Andreasen and Hein, 1987). On the basis of a sufficiently developed project brief and a consistent and clear set of criteria, a project approval decision may be made, including what priority it should have or whether it should be delayed. Notwithstanding the above, project briefs will vary in format, and can range from simple verbal requests to detailed documents. Also, within a project brief (or separate from it, or even in place of it), there could be a more narrowly defined design brief outlining the basic product requirements.

In contrast, if the design work is to be contracted out then a call for bids, an enquiry, or an invitation to tender will usually be issued. Depending on the organisation, and the scope of the work involved, this may vary from a basic verbal request to a written, and possibly very detailed, request for proposal or RFP (Hajek, 1977). If a decision to bid is made, then the response to this is a project proposal, tender, or bid.

Proposal requests may take different forms from industry to industry and from firm to firm. However, they should indicate the information necessary for the offeror to prepare a technical proposal and price quotation, and will normally specify the content and form of presentation required. The technical requirements, and thereby the latitude in what may be offered, may vary between being definitive or broad in scope. As bids are usually assessed according to criteria drawn from the technical proposal requirements, it is important to prepare a bid which is responsive to these (Hales, 1993), and to emphasise the appropriate competencies (Leech and Turner, 1990). The proposal is a written document which may be variable in length, but of a format which will generally cover details of the company's credentials; statement of the perceived problem; technical discussion; work to be carried out; project organisation, procedures and resources; project schedule; and cost breakdown (Hajek, 1977; Hales, 1993). This may include a technical synopsis, and indeed a detailed technical specification (or design brief) may be developed, although this is not necessarily for inclusion within the proposal.

It will be evident from the above that the project proposal (bid document) and project brief have some similarities. However, there are quantitative and qualitative differences as a consequence of their respective aims. Whereas a project brief is written with the aim of defining the development work to be carried out, the main aim of the project proposal is to secure a contract. In the first instance, the project proposal is used for bid selection and evaluation, then subsequently as a basis for negotiating contractual terms, where elements such as price, performance specification, work scope, and contractual terms may be subject to change.

Clearly, therefore, specifications serve a variety of purposes and hence there are a number of different types. Many sources, including the Cornfield Report (1979) and Walsh *et al.* (1992), have identified comprehensive specifications as being vital to success in product design. A comprehensive specification does not necessarily imply a detailed or prescriptive definition of requirements and constraints, but rather one where relevant factors are not omitted. Hence, the consideration and appropriate definition of all the key aspects is often regarded as an essential prerequisite to commencing the main project design activities. There are two important aspects to this: the specification of the product design requirements, and the more broadly defined specification of the project. The important generic features of these will now be discussed.

Design methodologists are in general agreement that it is necessary to establish a clear definition of the problem and a specification of requirements and constraints for which solutions will be sought as part of a design project. This is commonly referred to as clarification of the design task or objective. In practice, this will often require some preliminary design studies, and depending on the initial design strategy adopted may involve solution conjectures or abstractive analysis.

If an initial brief is provided this will usually state in outline the general objectives, requirements and constraints. However, this will not normally provide a proper basis from which to progress the main design stages, and it is therefore often recommended, as well as being common in practice, to produce a more comprehensive design brief or specification. In the different circumstances which occur in industry several names are applied to briefs and specifications; however, this working document is usually referred to as the product design specification, or PDS (Pugh, 1990), and is both formal and definitive. It is this document which defines all the requirements and constraints that have to be observed (British Standards Institution, 1989), by establishing succinct and precise performance requirements for each of the required attributes of the product (Cross, 1989).

The importance which has been accorded to the PDS cannot be overemphasised. Indeed, Pugh (1986b) has described it as 'the bedrock on which any competitive design must be based'. It is the basic reference source with which all involved in the product development interact and provides the main control for the product development activity. Consequently, the development and writing of the PDS is what Hollins and Pugh (1990) define as 'the most important part of the design process' (p. 84). However, contrary to this, they have observed 'the woeful inadequacy of the product design specifications in companies' (p. 86). Moreover, Walsh *et al.* (1992) suggest that it is common for many of the key elements, especially those relating to time and cost constraints, to be inadequately dealt with.

There is a broad agreement in the literature on what the basic content of a PDS should comprise. The goal is to establish a list of all the requirements and constraints that affect the design, as a failure to do this will result in a partial specification of the product (Pugh, 1990). To overcome this several authors (Pahl and Beitz, 1984; British Standards Institution, 1989; Pugh, 1990; Hollins and Pugh, 1990; Ullman, 1992; Hales, 1993) have proposed checklists, organised according to the type of requirement, for developing the PDS so that it is comprehensive, cohesive and unambiguous. However, Walsh *et al.* (1992) consider none of these to have universal applicability as much depends on the particular industry and product concerned. Moreover, Oakley (1984) has suggested that the form and detail of the specification depends on the complexity and scale of the project. Clearly, there is no single format for a PDS which is ideal in all situations, and indeed each reference source proposes a different set of detailed headings. There is, however, a general consensus that the main areas of concern must be with the product's performance, and with the time and cost constraints which apply. Some suggest grading the design elements, acknowledging that their relative importance will be influenced by the innovative or incremental nature of the product change (Pugh, 1990; Hollins and Pugh, 1990). Also, in developing an appropriately formulated and well-structured design specification, it may be useful to distinguish between those attributes or requirements that are demands, and those that are wishes. Lastly, as the PDS is a user document, it may be best written in a succinct and clear manner, using short definitive statements (Hollins and Pugh, 1990).

The requirements and constraints of the PDS establish the bounds of the potential solution space and thereby limit the range of acceptable solutions. Hence, the problem needs to be formulated in such a way as to leave the design team with an appropriate degree of freedom. In particular, many design methodologists stress the need for requirements to be stated in a way which is independent of any particular solution. Cross (1989) captures this when he states that 'the purpose of the specification is to define the required performance and not the product'. It should, therefore, not impose design solutions. However, several texts associated with the management of design or product development qualify this. In practice, briefs and specification are rarely produced without some idea of intended or possible design solutions, and successful design strategies may be based on a product-focused approach. Furthermore, the PDS itself may only be written following the creation and proof testing of feasible concepts (Walsh *et al.*, 1992), and there may well be legitimate reasons for using known principles, components, materials or designs (Andreasen and Hein, 1987). It may be concluded, therefore, that the PDS should not constrain the designer's choice of solutions unnecessarily.

However, achieving this balance may be difficult, as the statements of customers or clients are often couched in terms of solutions. There may also be confusion between the attributes of a product and its engineering characteristics. It is necessary to ensure that what the customer wants in terms of product attributes are carefully translated into specifications of the appropriate engineering characteristics (Cross, 1989). Traditionally this is undertaken by the marketing or engineering function, depending on the relative dominance between them. However the marketing function is unlikely to convey the technical issues adequately, and so create ambiguity, whilst the engineering department is likely to misinterpret or presuppose the market need. One approach, illustrated by Walsh *et al.* (1992), is to develop a detailed market specification from the project brief, which is then translated into a detailed technical specification by the design and production engineers. In achieving a feasible specification there will often be the need to qualify requirements and to acknowledge inevitable compromises. In ensuring that all the relevant design elements are addressed thoroughly, it is therefore important that all views are taken into account, and that the development of the PDS is regarded as a multifunctional task to be accomplished by the whole design team. Alternatively, the quality function deployment (QFD) method (see Chapter 8), which shares many of the desirable features already outlined above as well as drawing on other design methods, provides a more formal and comprehensive method for matching customer requirements to engineering characteristics. Most authors support the systematic preparation of design specifications, either based on the listing and ranking of requirements (performance specification method) or through more sophisticated and rigorous techniques such as QFD. Hales (1993), however, suggests that it may not always be necessary to compile a PDS in such a formal manner. Under certain circumstances the team may be so close to the problem that the requirements are clearly implied.

With design being a process which is iterative and interactive at the overall system, sub-system and component levels, Pugh (1990) suggests that it is useful to use this breakdown, and refer as appropriate to sub-system design specifications and component design specifications. Apart from the organisational logic in this, it is also appropriate since the relative importance and characteristics of the design requirements and constraints are likely to change at different levels in the product hierarchy. This is due to differences in the local environment for the product, sub-systems, and components respectively. To demonstrate this consider the motor car and the specification of operating temperatures. With regard to the vehicle as a whole, the ambient temperatures are clearly relevant; however for the design of components such as the engine pistons, the much higher temperatures within the engine cylinder must be considered.

Many design texts focus on the development of the PDS. However, as product development projects have much broader management concerns than this, it is necessary to provide some form of specification for the entire project which relates it to the commercial and strategic considerations of the business, and establishes what, and how, targets should be attained. Although in general this is referred to as the project brief (or project specification), in practice it may or may not be presented as a single document. Indeed in many cases the notion of a project file is probably closer to the truth. This document, regardless of form, should provide an appropriate description of the following:

- The project definition – outlining the project objectives and overall strategy; targets for different areas (including those relating to the product which are contained in the PDS); and the project plans, resourcing requirements, organisation, and procedures to be observed.

- The commercial and financial definition – outlining the potential sales and profitability, etc.; market characteristics, competition, and market strategy; product strategy, including the required product and cost structures; production strategy and costs; and a financial evaluation, including investments, liquidity and yield.

The context of design projects

The design practitioner's first task at the outset of any design project is to decide how to approach the problem. However, in using design models which prescribe a general approach to a broad group of problems, it is necessary for designers to interpret the models in the context of their own environment and the problem to hand. There are numerous factors, both internal and external to a company, which influence the requirements and characteristics of design projects. These are principally related to the characteristics of the company's organisation, and its products, markets, suppliers, production process, and the local and global environment. Unfortunately, by viewing the design process as a relatively autonomous and bounded activity, engineering design methodologists have tended not to capture or accommodate the consequences of these influences.

To some extent this has not been a particularly significant shortcoming of the design models in the past. Traditionally, the engineering design process has constituted one phase of a process which in overall terms was predominantly linear. The designer's role was to translate the given design specification into a detailed product definition. Other functional departments (purchasing, production engineering, production, sales) would in turn use this information to produce a manufactured product for sale or delivery to the customer. However, a direct consequence of the more rapidly changing nature of today's competitive environment is that an increasing number of companies are faced with greater market pressures. These often include demands for improved products, of higher quality, at reduced cost, and with shorter lead times. In these circumstances design increasingly has to be managed as a concurrent, multidisciplinary process.

This trend is in itself an overgeneralisation due to the different influences which exist from market to market, company to company, and project to project. Key success criteria include variables such as price, delivery lead time, delivery conformance, performance, quality and reliability, environmental factors, life cycle costs, and so on. However, each of these will be accorded different priorities between a company's products, markets and individual customers. Indeed, as indicated earlier, they impact on the company's strategic policies and hence the role of the design function. Is the company a product leader, developing products which excel in terms of the performance, quality and reliability? Or is the company an operationally effective manufacturer, in which the design role is concerned with minor improvements and minimising costs? Moreover, they impact both directly and through the strategic policies (such as types of markets to be targeted, types of product to be developed and the nature of changes involved) on a number of project characteristics. These can include the type of project organisation; the overall project and design processes, including the degree of process concurrency and extent of integration between functions and disciplines (as well as customers and suppliers); project planning and project management; and which methods and techniques are appropriate to use (Maffin, 1996).

Although there are many contextual factors, some of these can be considered relatively constant for any given company. The implication of this is that general approaches to projects, or dominant project modes, can be established for the principal project characteristics, provided the projects are within broadly similar product and market areas. Attempts to establish standardisation, in terms of the overall project approach, do not dictate the nature of the design process, but they inevitably limit the scope of the designer's role. It is important to realise, therefore, that through these efforts, and through establishing the project and product design specifications, considerable constraints are placed on designers. Care must be taken to ensure that the flexibility available to the designer is not excessively limited, for if it is the choice of design strategy, and the ability to conceive and develop satisfactory solutions to complex problems, will also be restricted.

Summary

In industry there has not been a widespread use of the problem-focused design strategy proposed by most engineering models. The design strategy most commonly pursued by design practitioners has principally been product-focused in nature. Consequently those models which have used descriptive studies as their basis are also predominantly product-focused. However, despite this apparent dichotomy, aspects of both approaches may be required in many design situations. In particular different strategies may be required for the various parts of a design project because of differences in their respective problem characteristics. Indeed, there is evidence that this is what occurs in practice, with design activity usually being characterised by a product-focused approach during the initial problem definition stages and aspects of both approaches being adopted during the main design stages (Maffin, 1996).

There are numerous contextual factors, both internal and external to a company, which influence the requirements and characteristics of a design project. By taking a holistic view of engineering design, it is apparent that a number of project characteristics are hierarchically related. These include the decision-making mechanisms associated with product strategy, project initiation and specification, and the execution of design projects. Unfortunately, by viewing engineering design as a relatively bounded process and attempting to promote a particular strategy, most engineering design models tend not to capture or accommodate the consequences of the contextual influences. To enable good design practice, it is not only important to be aware of these influences, but it is a necessity to understand their implications. This does not necessarily require new models as such, but that the current understanding of the design process needs to be developed within a broader context. This would allow individuals to make informed decisions in the context of their company environment and a particular project's requirements. It is only recently that work has been undertaken from such a perspective (Hales, 1993; Maffin, 1996); however, this does provide a logical basis from which to bridge the current dichotomy between theory and practice, and so enable a consensus to emerge between engineering design models, design teaching and design practice.

Chapter 7
Information in design

Introduction

The raw material of the design process is information, and therefore the designer's principal skill is one of information handling. The central position of information in defining the designer's work was shown in Chapter 1, where it was pointed out that the designer's office is filled with tools for handling information, such as calculators, computers, telephones, fax machines, drawing boards, manuals and catalogues. The many types of model used to define and evaluate a design, which were discussed in Chapter 4, also demonstrate the importance of information. Models are, after all, the physical manifestation of an otherwise abstract commodity: information. Designers use models (whether sketches, lists, equations, or computer software) to grasp the necessary information firmly, and to manipulate it to achieve their desired goals.

One view of problem solving suggests that it is the manipulation of information which provides the key in all cases (Simon, 1969). Relevant knowledge is repeatedly gathered and manipulated until the problem's solution becomes self-evident. This is clearly the case in algebra, where given equations are combined, reconfigured and reinterpreted (sometimes over many pages) until a solution is found. During this process deductive procedures may be used which could be said to generate new information, but it is clear from the bounded nature of an algebraic proof that the solution must result from the preliminary information being manipulated into an entirely new form. The same principle can be applied to other types of problem solving, including that of design. In this case the boundaries of the problem are not so clearly defined, and therefore the preliminary information includes all relevant knowledge. In practice the initial information contained in a project file will have to be augmented by retrieving additional data at intermediate stages, and this may be drawn from many sources including the client, manufacturer, regulatory authorities, manuals and theoretical texts. All this can be viewed as existing knowledge which the designer manipulates and transforms (on occasion creating additional information by deduction and induction) until, as with the algebraic problem, the solution is self-evident.

In this chapter the information processing activities undertaken by the designer are discussed and categorised. Each of these activities requires a certain amount of skill and training, but it is by using these skills that the designer can successfully find and define an arrangement of information which will provide a solution to a given design problem. However, during this process the designer will generate a vast volume of new information, and the sheer quantity of information can itself cause difficulties. The nature of this information explosion will also be discussed here, while in the next chapter the tools used to manage the flow of information are examined.

Information processing

Many activities process information, but it is possible to identify five categories of action that operate on information in distinct ways. Each of these will be discussed below, and the range of operations that make up each category detailed, but first it is useful to consider the relationship of each to the entire design process. The five categories of action that operate on information are:

- Collection.
- Transformation.
- Evaluation.
- Communication.
- Storage.

Although these activities have been arranged in a way which reflects the linear models of design, as discussed in Chapter 5, it should be recognised that in fact all these operations are being used at all stages of the design process. For example, data can be evaluated as it is collected, it is continually being communicated between members of the design team (and other interested parties not directly involved in the design process), and it must be stored and accessible at all times.

Clearly these activities are not called on sequentially, but used as and when necessary. A useful and illuminating comparison is with a computer, whose component parts are shown schematically in Figure 7.1. If the entire design process represents the computer, then the functions listed above can be placed in a similar configuration, as also shown in the same figure. Computers are of course information-processing machines, and so the success of this analogy is due to the fact that design can be considered an information-processing activity.

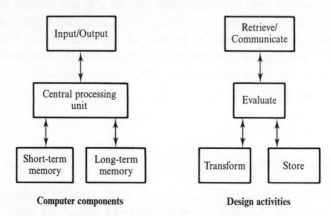

Figure 7.1 Information processing in computers and in design.

Information-processing activities

Data collection

Clearly the collection of information is important at the outset of a project, but as the design develops so further information will be required to enable each stage to proceed. The sources of information are varied, and this is reflected in the types of activity that can be grouped under this heading.

First, there is information *retrieval*, from published sources such as catalogues, manuals, journals, company reports, financial reports and so on. The designer may also have access to other unpublished documents such as marketing surveys, or government-sponsored reports. To gather information in this way successfully the designer must have good research skills.

A second form of information collection, that of *elicitation*, requires good interpersonal and social skills. In this case the designer talks directly to informed people, and by a process of careful questioning and even more careful listening tries to obtain pertinent information from the extensive, if unfocused, knowledge they have. Clearly the customer or client commissioning the design work will have many useful insights, as will the manufacturer. Other candidates for this approach are those who will have to sell or use the product. For made-to-order (i.e. one-off) items it is the operator who is most important, while for mass-produced products the marketing personnel are.

A third type of information collection can be termed *interpretation*. In many cases the designer is severely limited in the designs that can be produced due to the constraints imposed by codes of practice, and other regulations relevant to each specific industry. The designer therefore has to establish the limits within which any proposals must lie. Careful reading of the appropriate rules may be sufficient, but on occasion ambiguities will be found which require direct negotiation with the ruling body. These will establish how the rules should be applied in a given situation.

Information transformation

Design is a challenging activity, but it is also exciting, stimulating and rewarding. This is due in part to the fact that it is a creative process and, as with other exercises which produce entirely new entities, it is extremely satisfying when successful. In design the creation of something new is achieved by transforming information: knowledge about the way many things behave (which is gained from science and technology) is used to suggest an innovative arrangement of certain things that will behave in a desired manner. This is true of all design, but the point can be illustrated with one inspiring example: in California designers have created a powered aircraft which can stay airborne for at least twenty days, and which has successfully circumnavigated the world without landing or refuelling. This startling achievement is the result of combining knowledge of aerodynamics, lightweight materials and efficient power sources to produce an artefact with a performance that has never been possible before.

The creative transformations which are necessary to achieve such successful design were categorised by March (1976) into three types of reasoning, briefly summarised here. First, there is *productive* reasoning, during which knowledge of solutions to previous problems is used to conjecture a proposed design. Second, there is *deductive* reasoning, which uses known theories to predict the behaviour of a specific proposal, and finally there is *inductive* reasoning, which takes the results of a specific case to infer the likely behaviour of other possible proposals. March's model, which was presented in Chapter 5, links these three processes into an iterative cycle.

An alternative view of the way information is transformed during the design process is to recognise that all information manipulation is intended to achieve one of two objectives. These are either to elaborate or to simplify what is already known. In the process of elaboration greater detail is required in order to define some aspect of the design more precisely, or to establish the performance characteristics more accurately. Elaborative transformations are often required in the process of evaluating some aspects of a design. The effect of elaboration is always that the volume of information is increased. In contrast, simplifying transformations reduce the volume of information, as the purpose of these is to strip out unnecessary data, retaining only that which is essential for subsequent activities. Simplification is often advisable when aspects of a design are to be communicated, as the recipient will rarely need all the data previously generated.

Evaluation

If information is the material with which designers work, then clearly it is important that the designer monitors its quality. The information that is gathered at the start of the design process, the transitory information that is generated along the way, and the transformed information which is passed on as a final design all have to be evaluated to ensure that the quality is adequate. The quality of information can be assessed in two ways, these being its *accuracy*, and its *fitness for purpose*.

The degree of accuracy required is not a constant, but varies as the design process progresses. At the concept stage rough 'back of an envelope' calculations may well provide an adequate understanding, while at later stages more careful estimations are necessary. At the detailed stage rigorous analysis may be called for, but even then the calculating procedures need only provide sufficient accuracy for the requirement. To take two examples, the accuracy of the weight calculation for an aircraft is clearly of extreme importance, while that for a domestic appliance such as a video recorder is almost superfluous. In contrast the accuracy of the cost estimation may well decide the success or failure of both projects, even if it is for different reasons. In the former the initial outlay is large and can incur high interest charges over an extended period, while in the latter the profit margins may be extremely small.

With some types of information accuracy is hard to measure. Where opinions or value judgements are required, the range of answers given to a simple question may be wide. An example of this type of information can be found in design for safety, where it may only be possible to identify the best design if a precise value is placed on a human life in hard financial terms. In these cases clearly no one answer is more accurate than another, just as no single opinion can be considered right or wrong. Under such circumstances it is important to use several sources, canvass a number of opinions, and then establish the trend or average answer. Accuracy of this type of result is therefore dependent on the size of the survey undertaken.

The second aspect of evaluating information, ensuring that it is fit for purpose, is a major part of all design activity, and is necessary every time information is communicated within a design project. Consider any example of an information transaction which occurs within the design process, such as information provided by the client to the designer (the design brief); information for detailed analysis passed by the designer to specialist consultants; information about process capabilities provided by the manufacturer; or even the package of information provided by the designer to the client (the final design). It is self-evident that in every case adequate information must be provided to achieve the result required, i.e. it must be fit for purpose. In the last case, for instance, the information supplied must describe the product in adequate detail for the client to be sure that it meets the original specification, and for the manufacturer to build or produce it.

Communication

Before examining how information can be communicated, it is worth considering briefly what information should be communicated, as there are two potentially conflicting aims. On the one hand it is preferable to communicate only appropriate information, and so avoid overloading the recipient with unnecessary data, but this must be balanced against the additional work involved in providing each recipient with tailored information, rather than a standard set of data.

This point can be illustrated by looking at the final outputs of a design, and by way of example we can examine ship design. At the conclusion of the design project, several people require the designers to provide a precise definition of the vessel, but although all are interested in the same product, each requires a different definition. The naval architects must provide the client with a precise definition of the functional capability of the final design, and this will show such things as its accommodation arrangement for the given crew size, the engine-room layout to provide power for the given speed, the tank volumes for the given voyage duration, and the cargo holds and handling equipment for the given capacities and loading rates. But the naval architects must also provide a manufacturing definition to the shipyard which will build the ship, and this will include different information such as the precise geometry of the hull, the specification of the steel to be used in construction, the position of the welds in the plating, the routing of pipework between tanks and pumps and the sources of standardised components. In addition to this the designers may have to provide further definitions for other parties. For example, the operators may require handbooks detailing the capabilities of specific aspects of the design (such as a stability booklet for the ship) and maintenance procedures and schedules; and the regulators or approval authorities may require evidence proving that the design is within their requirements, possibly in the form of detailed calculations.

In the case above, four possible recipients of a design definition were mentioned – the client, the manufacturer, the operator, and the regulator. Each has a requirement for different information, but there will also be some considerable overlap. Providing all these parties with all possible design information would result in considerable confusion and incorrect decisions; but similarly tailoring information packages (including drawings) to suit each recipient's needs would entail considerable extra work. One compromise solution is to identify a body of information which can be considered core data, and provide this to all recipients. The designer then decides exactly what additional information is required by each recipient, tailoring this part of the information package to their specific requirements.

In the above discussion drawings have been mentioned as one medium by which information can be communicated, but it is worth considering in greater detail the several mechanisms by which design data is transferred. Clearly, at the earliest stages of the design process iconic models such as simple sketches can communicate a concept proposal effectively, just as detailed scale drawings can provide precise information with great clarity at the later stages. Today both early concepts and detailed designs may have been generated on a computer, but the information may still be transferred in the form of hard copy printouts of specific views of the computer model. With many aspects of the design process now being performed with the aid of design software, electronic links may be in place which allow direct transfer of the data from one designer to another. This can extend to the manufacturing process, with numerically controlled machines being fed by links from the designer's database. Even when hard-wired links are not established, the transfer of information can be achieved down the ordinary telephone lines with the aid of modems, or alternatively by the physical medium of a floppy disk containing the required data.

Sketches and drawings are an effective means of communicating information concerning the relationships between elements, although these need be not only of a geometric nature, as demonstrated by circuit diagrams and other schematics of complex systems. While diagrams can communicate much, they often have to be complemented with written specifications which provide information not necessarily detailed on the drawing, such as the characteristics of each component. This supplementary data can include the material of manufacture, relevant tolerances and type of finish, or alternatively the detailed specification of a standard part in terms of operating performance or supplier's catalogue number.

Despite the old adage that 'a picture is worth a thousand words', not all types of information can be communicated in this way. In particular the justification of decisions requires a reasoned argument that is best set down in the form of a report. This may be highly technical in nature, and draw heavily on detailed analysis which can be explained only with supporting mathematical formulae. For designers the ability to communicate in this way is important, both to explain the basis of a proposed design and to elaborate on the more detailed argument which led to the final result. The client and regulators will require this information, but others working in the same field will also be interested if commercial interests do not prevent such information being published.

Lastly, consideration must be given to the transfer of information by verbal means. The ephemeral nature of oral communication suggests that it is always an inferior mechanism in comparison with a written report. This is not the case, however, and there are good reasons for communicating some information live, whether in the form of a formal presentation, or in an informal one-to-one discussion. The value of verbal exchanges is primarily in subtleties that are difficult to put into written reports. Authors may be reluctant to express opinions, value judgements and enthusiasm in a written report, but almost unintentionally include all of these when communicating verbally. Clearly if the intention is to persuade the client of the advantage (or disadvantage) of a particular decision, communicating enthusiasm (or disapproval) may be as important as the reasoned argument. In addition it should be recognised that with a verbal presentation the transfer of information is always in two directions. Although generally only one side is proactive in that they are making the presentation (either formally or informally), the recipients provide a great deal of reactive feedback even if it is only in the form of body language. At the earliest stages of a design project it may be just this sort of information which provides the designer with clues to the client's unconscious preferences and subtleties in the requirements or desired outcomes of the project.

The education of engineers has conventionally concentrated upon the development of analytical skills, encouraging a systematic methodical approach to problem solving in general, and to the activity of design in particular. In the past little emphasis was placed on the ability to communicate in ordinary language, considerably higher value being placed on numeracy than literacy, in both selection and assessment procedures. Nowadays, however, in what has been termed the information age, the importance of being able to communicate technical results effectively in an appropriate form for any given audience is widely recognised, and engineering courses are required to include in their syllabi instruction and practice in the preparation and presentation of reports, both written and oral.

Storage

For information to be of any value it must be accessible, therefore effective means of storing it must be established which will enable its efficient retrieval. Once again, paper-based storage facilities such as card indexes and filing cabinets are being superseded by computer databases and files. If these are properly organised the ability to locate and retrieve information is greatly enhanced. Care must be taken, however, to ensure adequate security from both unauthorised access and accidental loss by introducing effective password protocols and methodical data backup facilities. (It is interesting to note that a survey carried out in 1993 found that 75 per cent of designers still used paper-based, or hard copy storage, as well as any computer based system (Court *et al.*, 1993).)

The information that has to be stored can be categorised into three types of data: input, in-progress, and output. The first of these is the information which is used as the raw material of the design process.

Much of this information is contained in publications such as catalogues, manuals, codes of practice, and regulations, and as such is stored on the designer's own bookshelves or in specialist libraries. It is becoming common practice, however, for many of these to be distributed in the form of compact discs (CD ROMs) which are accessed via a computer. To this type of background information the designer adds information which is specific to the project, and which is gathered from several sources at the outset of the design process. Clearly this includes all the ideas and requirements that the client supplies, and the results of market surveys, but it also includes a trawl through previous work of a similar nature which might provide useful insights, or even a base from which to launch the new work. Much of the background information which is gathered will prove to be irrelevant, but only as it is filtered and evaluated will the importance of some elements become apparent, and these will be the building blocks of the new design.

The second type of information that has to be stored is information which is created during the design process: the in-progress data. As design decisions are made they must be recorded so that the progress of the project is maintained and completed areas are not reworked unnecessarily. For small projects the maintenance of a log book may be an adequate way of ensuring that the development of the design is properly recorded, and mistakes are not made due to misunderstandings over which aspects of the design are finalised and which are still in a fluid state. On a larger project regular reports may be required which detail the progress in particular areas. These are used by the project management team to co-ordinate the entire operation, and to ensure that each design group is aware of the relevant elements which have been finalised. Clearly these reports are also used for other aspects of the management of the design process, such as ensuring that design resources are allocated appropriately, and so ensuring that those areas which are encountering difficulties are brought back onto schedule.

The last type of data which must be stored is the output of the design process. The nature of the information which is passed out of the design office is discussed below, but there are two reasons why it is important that the design office itself maintains a record of all information gathered and generated during the process. These are that this information represents both an insurance policy and an investment portfolio.

Completing a design project does not necessarily terminate the designer's involvement with the product. Some level of design support will be required throughout the product's life cycle, and as a result any of the details generated during the design process may be required at a future date. This could be for a variety of reasons, such as allowing alterations to be made after some years in service, to respond to questions regarding maintenance, or to establish if alternative modes of operation are safe. Especially when a failure has occurred, the ability of the designers to establish both the cause and the responsibility may be crucial. An example of this can be drawn from the offshore industry where, in 1993, a completed concrete oil production platform undergoing ballasting tests in a Norwegian fjord suffered a catastrophic failure in one tank (*Offshore Engineer*, 1991). The platform sank in minutes, breaking up on the sea bed, with a resulting replacement cost of many millions of pounds. A review of the design drawings and calculations was necessary to establish if this failure was due to poor manufacturing procedures or to a fundamental inadequacy in the design itself.

Clearly then, stored design information can be considered as an insurance policy against unjustified litigation, but its value as an investment must also be recognised. In the cases of variant and adaptive design, having detailed information on previous products to use as a basis for subsequent designs is clearly an advantage. Even in the case of original design the calculations and analysis carried out in earlier work may well provide useful insights, informing the designer of profitable concepts to explore. Keeping a careful record of previous design work is therefore important for every designer and design company, as this information is not possessed by any other design office. It is this corporate knowledge which provides a company with an advantage when bidding for future contracts of a similar nature, as much of the groundwork has already been undertaken.

The information explosion

As a design progresses the product is defined in greater and greater detail. In addition each aspect of the design is evaluated with analysis procedures of increasing complexity and accuracy. An inevitable result of this is that the quantity of information generated grows exponentially, increasing the difficulties of finding the correct information when needed and of communicating appropriate information to the relevant interested parties.

This expansion of information is compounded further when, in response to pressure to accelerate the design process, aspects of the design which would conventionally have been considered sequentially are addressed concurrently, this being termed *simultaneous engineering*. In principle this approach can be implemented wherever multiple requirements have to be addressed. These could be within the design process itself, such as the requirements introduced by functionality and producibility, or between the design process and other related activities, such as production process design, marketing activities, and supply or distribution preparations.

An inevitable result of addressing multiple requirements simultaneously is that not only does the quantity of information generated grow explosively, but also it increases the difficulties of finding the correct information when needed, and of communicating appropriate information to the relevant interested parties. Clearly, in this case, the management of the flow of information becomes a highly sophisticated activity. The question as to what information is required by each design group must be carefully identified, but in addition it must be established just what is the minimum information that will enable each team to embark on their particular part of the design process, even though it is accepted that further input will be needed before they can complete their task. This concept of *partial* information (i.e. incomplete, but sufficient to make a start) has to be linked with that of a critical path analysis for the generation of information, and the making of design decisions. The flow of information is therefore carefully planned out so that each team can start on their task as early as possible with a minimum of information, receiving further data as it becomes available, and passing on their results and conclusions in time for other teams to complete their own tasks. It should be recognised, however, that this strategy involves added risks: should decisions have to be changed the ramifications are potentially more far reaching than with a sequential approach to design.

Containing the explosion

While the rapid growth of information during the design process is unavoidable, the designer must keep it under control and, in the final stages of the process, reverse the trend so that the volume of information is reduced. The final recipients of the design do not require all the data that has been generated along the way but only the conclusions, and these presented in as simple a form as possible.

For some participants in the design process it is possible to avoid the accelerating growth in information during the process, and therefore to avoid the major implications for information storage, retrieval and management. In the case of highly complex items, such as oil production platforms, the large multinational companies planning to operate them may undertake the concept design work in-house, then subcontract out the more detailed stages of the design. In this way they are only working with the relatively small amounts of data required in the early stages of the project, and then receiving the condensed data comprising the final design at the end. The complex problem of managing the vast amount of intermediary data generated by the detailed (and multidisciplinary) analysis of many aspects of the design is dealt with entirely by the subcontractors. Designers working on a smaller scale can also avoid unnecessary information generation if they are aware of the capabilities of the recipients of their designs. An experienced manufacturer will require less design data than a manufacturer who is being asked to produce an item for the first time, as details which are understood implicitly by the former may have to be made explicit for the latter.

Strategies such as those described above, however, do not prevent the information explosion; they simply transfer its impact to alternative organisations. Clearly, within these alternative organisations there will also be designers, and it is these who have to actively confront the difficulties resulting from the volume of information generated. A general principal of design (elevated to one of the two 'axioms of design' by Suh (1990)) is that information should be kept to a minimum throughout the process. There are essentially two techniques for achieving this, by increasing repetition or by reducing accuracy, both of which deserve further comment.

First, the amount of information required to define a complex product can be reduced dramatically if there is an element of repetition. Instead of having to define each element individually, only one needs to be detailed, with the second being specified as a repeat of the first. This concept can be observed in several practices commonly adopted by designers:

- Symmetry: axes of symmetry result in half the artefact being duplicated to form the whole. This may reduce by almost a half the necessary descriptive information. Take, for example, a simple office desk: if an identical set of drawers is to be provided on each side of a central sitting position the definition will be considerably simpler than one with different storage arrangements to the left and right. It is also likely to be simpler than the definition of a desk with drawers provided only on one side, as this also has no axis of symmetry.

- Industry standards: standards such as those published by the British Standards Institute can reduce the volume of information required to specify a design. This is because the single statement indicating that the item should conform to a precise standard communicates a large body of information, without having to repeat it explicitly in the specification.

- Standardised modules: using standard modules or components repeatedly in a design clearly simplifies the product definition.

- Variant design: a complex product can be defined by reference to a previous design, indicating that the new product is almost identical, the specific points of variance being itemised and defined in detail.

All such practices reduce the total quantity of information required, as instead of defining every component in detail they simply point to the relevant template, whether it be part of the design package or external to it. This principle can lead to surprising results, such as the foot pedal on a car with two lugs on it to attach a return spring, even though only one is ever used. Providing the additional lug allows the same pedal to be installed in cars with either left- or right-hand drive, and so removes the need to design, make or stock an additional component.

The second way to control the volume of information is to reduce the accuracy of the information generated. Simpler analysis and evaluation procedures are less onerous as they require less input data and involve less calculation (and therefore generate less intermediate data), but the results they produce have a lower level of accuracy. The designer must balance the resources available with the accuracy that is required. A lower level of accuracy implies an increased level of risk, and this increased risk is either tolerated or compensated for. In marketing surveys or costing calculations quick results can be obtained but only with an increased probability of error. This risk may be acceptable if the potential reward is high, as in the case where a promptly submitted proposal is most likely to win a seemingly lucrative contract. On the other hand, in structural design the consequences of failure can lead to injury or even death; therefore low accuracy must be compensated for by increasing the factor of safety. All structures are designed to withstand a greater load than that anticipated, but if rough-and-ready evaluation methods are used, this factor is high, whereas if detailed analysis is carried out, only marginal factors are needed. In aircraft design, for example, where the penalty for excess weight is severe, very rigorous methods (such as finite element analysis) are combined with low safety factors. In contrast civil engineers often have little or no penalty to pay for unnecessary weight, so simpler and quicker empirical methods can be used, together with high factors of safety. These examples explain why, in certain quarters, factors of safety are sometimes referred to as 'factors of ignorance'.

Summary

The medium in which the designer works is that of information, and the skills required by the designer are those of information handling. The ability to collect, evaluate, manipulate and communicate many types of information enables the designer to find a design solution which satisfies the client's requirements. Today, however, the designer requires an additional skill, that of managing the flow of information. Pressures from industry dictate that the design process be carried out more rapidly, and from society that consideration be given to many more requirements than simply that of functionality. As a result the design process involves a complex net of participants, all of whom require and generate elements of design information. Ensuring that the process is not swamped by the sheer volume of data, nor delayed by poor scheduling of design activities, necessitates that the information flow in every design project is managed effectively.

Chapter 8
Tools for design

Introduction

This, the last chapter of the book, describes some of the tools, techniques, and methods available to designers for the purposes of retrieving, generating, manipulating, transforming, evaluating and communicating information. Some of the tools are generic and can be used for a variety of purposes in a wide range of disciplines. Other tools are discipline-centred and have no potential outside that discipline. In any one discipline, however, the sum total of the tools, techniques and methods available to a designer represents a symbolic tool kit.

The tool kit presented in this chapter is intended to provide a representative sample from the wide range of tools available. These could be categorised in a number of ways, such as the stage of the design process where they are applied or according to their usefulness in aiding convergent or divergent thinking. However, it is worth noting that some tools may be used at a number of stages throughout the design process and are not confined to any particular one. The approach adopted here therefore focuses on the nature of the tools themselves, rather than on their use or application. Thus, the tool kit is presented in three sections: information-based tools, procedure-based tools and computer-based tools.

Information-based tools

Information-based tools essentially fall into two categories. The first category comprises the information resources that are available to a designer and which can be used in a variety of ways, such as providing the basis for a new project, stimulating innovative thinking or as a guide to legitimate or compliant solutions during the process of solving a problem. The second category contains the more activity-based tools that can be used, say, to communicate information. Information-based tools are key components of a designer's tool kit, as the discussion in the last chapter emphasised.

Product catalogues and design handbooks

Product catalogues and design handbooks are an important and frequently used source of information for designers. Product catalogues contain information about the functionality of a component or product, details of its attributes, and usually photographs or pictures to give the designer an indication of its appearance. Design handbooks, on the other hand, provide a useful means of quickly bolstering a designer's knowledge about a particular concept or product. Many designers maintain an up-to-date library of such handbooks, and many manufacturers provide them for free because they want designers to select their products. Collecting design handbooks is therefore an easy way for newly qualified designers entering a particular industry to build up their own database of available resources.

It is worth noting that product catalogues and other technical reference material can be used to stimulate ideas for solving a particular problem. Many good ideas are publicised in design catalogues, trade journals, engineering institution publications, engineering magazines and so forth, and so every designer's effectiveness can be enhanced by monitoring these regularly. Indeed, it is the case that significant amounts of design effort can be eliminated from the entire design and manufacturing process by reapplication of existing and available technology. Even if this technology is patented, many organisations will grant licences whereby another company can use the technology on payment of a prearranged fee. What is most useful about reapplying existing and available technology is that there is a concomitant reduction in the risks associated with its use, since the ideas and concepts are well proven.

Standards, rules and regulations

Standards, rules and regulations can be issued by international organisations, governmental bodies, or independent classifications societies, respective examples of which are the International Maritime Organisation, the American National Standards Institution and Lloyds Register of Shipping. Such standards are not only necessary, but also useful as they can guide the design effort in safe and sensible directions. Standards and codes of practice delineate what is regarded as good engineering practice.

Standards that are important to design projects fall into three categories (Ullman, 1992):

- Performance standards – targets which must be achieved for a particular product to function correctly.

- Test method standards – for measuring the properties or attributes of a product that might need to be communicated to other parties, such as the users of that product.

- Codes of practice – which provide parameterised design guides for particular aspects of a product, such as the sizing of pressure vessels and so on.

The purpose of the various bodies in issuing standards is to ensure that designers create products that are safe and workable. The rules and regulations that constitute a standard result from many years of experience in addition to an understanding of the theory underlying a particular area of technology. Following them will therefore result in a product that is acceptable to the relevant authorities and will function correctly. Unfortunately, however, there can be disadvantages in using standards since a product which has been designed to the appropriate standard may be conventional, even mundane, despite the fact that it functions correctly and performs safely. The reason for this is that the standards embody conventional wisdom and reflect routine practice. Accordingly, it is difficult to conceive an innovative product if the approach adopted is 'design by rule'.

A more creative approach is to design from first principles. In this approach, designers use their own knowledge and in-depth understanding of the problem to conceive solutions which are theoretically sound, but which would not have been possible if designing by rule. Successful innovations eventually become embedded into the new standards and therefore become conventional themselves. The important thing about innovative design, however, is that no matter how radical an idea or solution is, the designer still has to satisfy the relevant bodies that it represents sound engineering practice. These institutions are open to new ideas, provided the designer can demonstrate that the idea will work in practice in a safe, reliable and effective manner.

A useful example which serves to demonstrate this point clearly can be taken from the marine industry. For many years ships had been designed and built using a transverse framing approach whereby the principal structural support consisted of horizontally-oriented stiffeners (i.e. they ran from one side of a vessel to the other). Towards the end of the nineteenth century, however, designers saw an opportunity to alter this configuration such that these principal structural members ran from the bow of a vessel to the stern, that is, from forward to aft. The commercial justification for this change in design philosophy was unquestionable, as much larger vessels could be built using the new approach. However, at that time all the regulations pertaining to naval architecture promoted the notion that transverse framing represented sound engineering practice. In order to satisfy the relevant regulatory bodies, designers had to demonstrate from first principles that the concept of longitudinal framing was valid from a technical point of view. Needless to say these efforts were successful and have led to many of the ships designed subsequently being longitudinally as opposed to transversely framed.

Historical product data

The previous chapter went to some length to demonstrate that an organisation's library of design solutions is actually a corporate asset which must be used for competitive advantage. Even when faced with an original design problem, a designer often tends to bring historical product information and solutions to bear. As was discussed in Chapters 5 and 6, many design problems are ill-structured in nature, and historical information and data can frequently provide the necessary catalyst to initiate a productive line of reasoning to help solve such problems. By gaining an insight into how a problem has been solved in the past many designers find that a whole host of alternative solutions emerge, one of which might prove to be a means of successfully solving that problem.

Furthermore, in cases of variant and adaptive design, historical knowledge is used in order to minimise the effort required to produce a successful solution to a particular problem. This approach has several advantages. First, the designer can choose to vary or adapt an existing design which is known to have been successful because it performed well, was economic to produce and so on. Second, the designer has the opportunity to address any known shortcomings with the selected base design so that the new design represents an improvement over the older one. It should be noted that in using existing designs as the basis of a solution to a new problem a designer is not stifling creativity. By using existing solutions in some areas, the designer is able to focus on innovative solutions for other aspects of the problem, secure in the knowledge that some pieces of the jigsaw have already been sorted and put into place.

As historical product data are proprietary knowledge (and possibly patented) which a company has paid for over a period of time, it is important to have a good cataloguing system which affords easy identification and retrieval of information from the design library. Additionally, it is important to keep this library up-to-date with new information as it becomes available, including data on the performance of a product in service or operation.

Sketches and drawings

Sketches and drawings are probably the most useful tool that a designer can employ to communicate design intent. The degree of detail included in a sketch or drawing can range from a crude outline to a precisely draughted engineering definition which would be used for manufacturing purposes. The effectiveness of sketches and drawings over other forms of communication is expressed in the old adage that 'a picture is worth a thousand words'.

Sketches and drawings are used throughout the design process. At the concept stage, when generating alternative solutions to a problem, sketches are often used to explore ideas and subsequently to present a proposal which can then be commented upon, or criticised, by others. Indeed, it is often said in design circles that an innovative idea was sparked off by a designer doing a sketch 'on the back of an envelope'. Such sketches provide the inspiration for a design team to suggest amendments or improvements to the initial suggestion, and so swiftly transform it from a crude idea into a viable concept. Conversely, at the detailing stage of design, engineering departments issue drawings to the relevant manufacturing departments both to communicate the form of the design and to provide explicit instructions which enable it to be manufactured. Unlike concept sketches however, these drawings are used to provide the detailed description of a product, or a particular part of that product.

As has been discussed in Chapter 4, sketches and drawings are one form of model. Clearly, models can be considered an extension of drawings as a design tool.

Procedure-based tools

Procedure-based tools are principally used for generating and evaluating information. These tools provide formalised techniques which ensure that specific tasks in the design process are tackled rigorously. Although such tools are prescriptive in their approach, they do not represent a deterministic mechanism for delivering results, but simply provide a framework for structuring and focusing designers' thought processes.

Quality function deployment

The principles and requirements for developing good specifications were discussed in Chapter 6. Quality function deployment, or QFD, is a rigorous technique which ensures that the important aspects of developing good specifications are adhered to. The aim of QFD is to establish targets for the engineering characteristics of a product, in order that a customer's requirements are subsequently satisfied. According to Cross (1989), the general procedure to be observed is as follows:

1. Identify customer requirements in terms of product attributes.
2. Determine the relative importance of the attributes.
3. Evaluate the attributes of competing products.
4. Draw a matrix of product attributes against engineering characteristics.
5. Identify the relationships between engineering characteristics and product attributes.
6. Identify any relevant interactions between engineering characteristics.
7. Set target figures to be achieved for the engineering characteristics.

QFD may be applied to the design of both new and existing products (Ullman, 1992). As it is a technique for identifying the features of products that are of value to customers, it may also be employed to identify the critical characteristics of existing products. It can be applied to the design of a complete system or some simple component part.

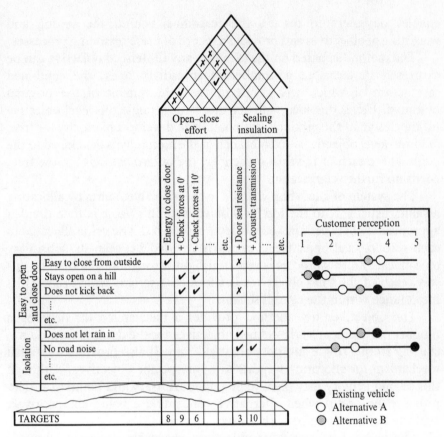

Figure 8.1 An example of QFD.
This diagram shows part of the analysis for a car door.

By virtue of its rigour it encourages the designer to understand the engineering characteristics of the problem and to think in detail about the design requirements. QFD is an adaptable technique which can be applied equally well during the later stages of the design process as it can in the earlier stages. A diagrammatic example is shown in Figure 8.1.

Objectives tree

Defining measurable objectives for the design of a product is a crucial task. Without a set of clearly defined objectives the processes of concept generation and evaluation will be unfocused. The objectives tree method (Cross, 1989) can be used to identify specific objectives, to illustrate the relationships between them and even to allocate weights to them. It enables designers to discuss the reasoning behind the setting and weighting of objectives and provides a record of these reasoning processes.

The method is based on the idea that any ill-defined objective can be expressed in terms of a number of subsidiary ones, the combined satisfaction of which will result in the achievement of the original objective. Using this idea, and starting with a single top-level objective for the design of the product, it is possible to develop a hierarchy (or tree) of lower-level objectives. Each branch of the hierarchy is developed to the point where each of the final objectives, the outermost twigs of the tree, needs no further clarification.

The system of allocating weights to the objectives starts by allocating a value, usually 1, to the top-level objective. This value is then divided appropriately between the second-level objectives. The value allocated to each second-level objective is similarly divided between its subsidiary objectives, and this process is repeated right throughout the tree. In this way, each of the final objectives is allocated a weight which reflects its importance within the overall scheme.

The objectives tree method provides a mechanism for deriving a number of specific, clearly defined objectives for the design of a product, starting from a single and possibly vague aim. It also provides a rational mechanism for allocating weights to these specific objectives. Although this second activity may seem to be rather arbitrary and pointless it does prove to be useful when alternative concepts are being competitively evaluated at a later stage of the design process.

Figure 8.2 shows part of an objectives tree for the design of a 2-metre household step-ladder. It clearly shows how the specific subsidiary objectives of 'low weight' and 'folds flat' are related to the vague top-level objective of 'overall convenience'. It also illustrates a practical mechanism for allocating weights to the subsidiary objectives. At each branching point, 'relative weights' are allocated to the subsidiary objectives such that the sum of these weights is equal to 1. The 'absolute weight' of each subsidiary objective can then be calculated by multiplying its relative weight by the absolute weight of its parent objective.

Functional analysis

The task of generating conceptual solutions to a novel design problem can be difficult. However, there are techniques whose purpose is to structure the process and divide it into smaller steps. Functional analysis is one such technique. It is used to split the design problem into a number

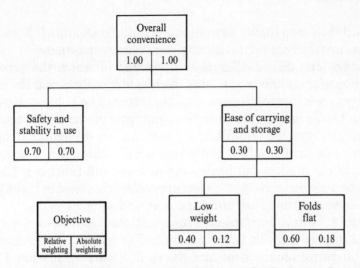

Figure 8.2 An example of an objectives tree, in this case for the design of a step-ladder. The objectives tree is used to break down the top-level objective into a number of subsidiary objectives. A system of weighting values can then be used to define the relative importance of all of the bottom-level objectives.

of smaller problems, the combined solution of which will result in a solution to the original problem.

In many design problems, the function of the product can be expressed in terms of 'inputs' and 'outputs', some of which are required while others are simply available utilities such as power supplies or waste disposal facilities. These inputs and outputs generally correspond to flows of materials, energy and information (i.e. signals). Block diagram notation can be used to express the overall function in these problems, and to explore the way, or ways, in which this function can be broken down into simpler sub-functions (Cross, 1989). Flows of materials, energy and information can be denoted on these diagrams using different line types.

At the outset it is important to ensure that the product's function is defined as broadly as possible within the scope of the specification or design brief. This is in order to avoid overlooking any functions which need to be incorporated into the product, or restricting unnecessarily the range of possible design solutions. Once this is done, it should be possible to draw a block diagram of the product as a system, showing all its inputs and outputs.

The next step is to decide how the overall function of the product can be broken down into sub-functions, recognising that it may be possible to do this in several ways. Sub-functions should be given short, snappy titles, and their own inputs and outputs should be identified. Some sub-functions may correspond to existing components or systems.

Having done this, a diagram is developed in which the product's system boundary is drawn as a large, rectangular outline, and the inputs and outputs are shown crossing this boundary. The sub-functions are drawn as blocks within this boundary, and their inputs and outputs are connected so that the sub-functions 'add up' to the required overall function. The final result is a diagram which shows how the overall function of the product can be decomposed into sub-functions. Clearly, this can be done recursively, the same approach being used to break down the sub-functions, and their sub-functions, and so on.

Figure 8.3 shows three stages of functional analysis for a coin-operated 'pay and display' parking ticket machine. The machine's user interface and maintenance requirements are shown in Figure 8.3a, while Figure 8.3b shows how the function of the machine might be broken down into four sub-functions. Figure 8.3c takes the analysis one step further, showing how one sub-function – the 'coin management system' – might itself be decomposed.

Brainstorming

Generating ideas which may provide solution concepts is a crucial activity in design. Brainstorming is a technique for generating ideas which makes use of the creative interactions within small groups of people. The process is straightforward – five or six people are gathered together and given a problem statement. After a few minutes during which they can gather their thoughts, they are asked to suggest possible solutions. All suggestions are recorded, and no criticism or evaluation of suggestions is allowed. A time limit for the session may be set, or alternatively it can be ended when there are no more suggestions.

This technique encourages people to make suggestions, no matter how silly they may seem, by removing the prospect of criticism and judgement. Once shared, an idea can be improved on by other members of the group, or can trigger other, related, ideas. Indeed many notable organisations, particularly in the product design field, regularly hold concept or brainstorming sessions whereby all sorts of ideas are accepted, at least initially. Tefal, Black & Decker, 3M and Sony are only some of the many companies who encourage their employees to get involved in these meetings. Interestingly the formal qualifications of the participants are not necessarily important, as it may be life experiences and the ability

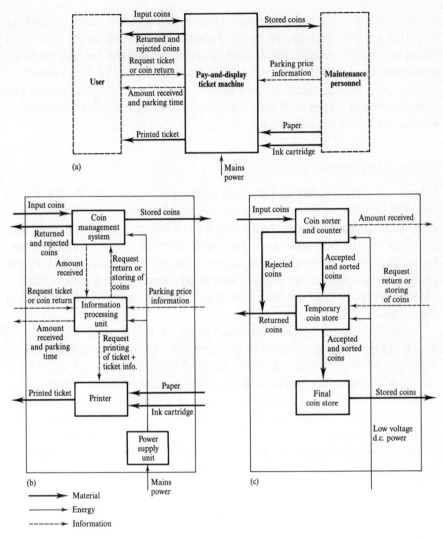

Figure 8.3 a, b, and c An example of functional analysis: a pay-and-display ticket vending machine.
Block diagrams can be used to break down the overall function of a system into a number of sub-functions. Each sub-function can then be analysed using the same approach.

to think laterally that ensure useful contributions are made to the brainstorming sessions. While many companies do use their own personnel, some rely totally on the suggestions from their customers to conjure up the next idea for a new product, or for an innovation to an existing one.

A brainstorming session should be followed by a subsequent meeting in which all the ideas raised in the brainstorming session are examined critically and thoroughly evaluated, possibly using the formal methods discussed later in this chapter. At this stage many of the ideas recorded during the brainstorming session will be discarded, but the session will have been a success even if only one or two of the ideas are considered to be worth further investigation.

Concept generation matrix

When the design problem has been decomposed into sub-problems, as described earlier, solutions to these sub-problems have to be generated. It is often desirable to generate a number of solution ideas for each sub-problem, as this provides scope for choice later on in the design process. In contrast to the somewhat anarchic approach of brainstorming, concept generation matrices (alternatively known as morphological charts (Cross, 1989; Ullman, 1992)) provide a formal structure for making and recording these choices.

The basic approach is straightforward. A matrix is constructed in which each row corresponds to one of the sub-problems. Brief descriptions or pictorial representations of the solution ideas relating to each sub-problem are inserted in the first few cells of the corresponding row. Thus, the finished matrix consists of a number of rows, each of different length, containing descriptions of all the solution ideas for all the sub-problems.

With the relevant information displayed in this way, the compatibility of the various solution ideas can be considered. The matrix may reveal pairs or larger groups of solutions which are well-suited to each other, perhaps because they make use of the same energy supply or because their characteristics tend to compensate for each other with varying ambient conditions. Conversely there will probably be pairs of sub-solutions which are incompatible. It should therefore be possible to identify a few well-suited combinations of solution ideas and to justify their choice on a rational basis. Each chosen combination can be represented on the matrix itself by a line joining together the selected sub-solutions.

The concept generation matrix is used to select favoured concepts from the large set which comprises all the possible combinations of sub-problem solutions. By showing the various solutions to all the sub-problems in a compact format, it helps the designer to focus on the compatibility or incompatibility of specific solution ideas. Figure 8.4

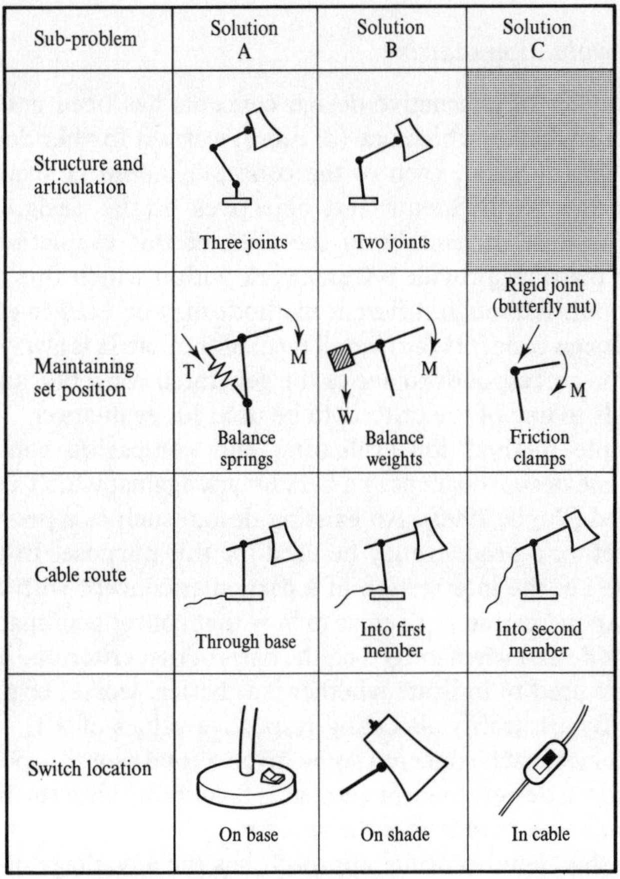

Sub-problem	Solution A	Solution B	Solution C
Structure and articulation	Three joints	Two joints	
Maintaining set position	Balance springs	Balance weights	Rigid joint (butterfly nut) / Friction clamps
Cable route	Through base	Into first member	Into second member
Switch location	On base	On shade	In cable

Figure 8.4   An example of a concept generation matrix, used here to explore possible designs for a desk lamp.

Concept generation matrices can be used to highlight affinities and incompatibilities between alternative sub-problem solutions. Concepts are generated by combining compatible sub-solutions.

shows a concept generation matrix for the design of an articulated desk lamp. It provides a useful illustration of how these matrices can be used to highlight affinities between solution ideas. In this case, inspection of the matrix reveals that the switch can only be located on the base if the cable is routed through the base, and that this cable route is simplified if the 'two joint' structure is used. Add the idea of balance weights to maintain the set position, and a complete concept has emerged.

Concept evaluation matrix

Once a number of alternative design concepts has been generated, the designer must decide which one (or more) warrant further development. To make this decision, each of the concepts should be evaluated with respect to the requirements and objectives of the design, and then compared against each other in the light of this evaluation. Concept evaluation matrices provide a framework within which this activity can be carried out. Although different methods may be used to evaluate and compare design concepts the overall form of the matrix is always the same: each column corresponds to one of the generated concepts, and each row corresponds to one of the criteria to be used for evaluation.

A simple method for evaluating and comparing concepts is to designate one design concept as a benchmark against which all the others are matched (Pugh, 1990). An existing design such as a previous model or a competitor's product may be used for this purpose. In each cell of the matrix (i.e. the intersection of a particular concept with a particular criterion) an indication is given as to how that concept compares with the benchmark design when judged on the basis of that criterion. Appropriate symbols are used to indicate whether it is better, worse, or pretty much the same. By arbitrarily allocating respective values of $+1$, -1 and 0 to these outcomes, each concept can be given a total score. A positive score indicates that a design concept compares favourably with the benchmark, and a negative score indicates the reverse.

While this 'point-scoring' approach has the advantage of simplicity, it is rather crude. A more sophisticated method (Cross, 1989) requires three separate stages. The first stage involves identifying a 'performance parameter' corresponding to each of the evaluation criteria. This parameter may have objective numerical values, or it may be a subjective scale, ranging perhaps from 'excellent' to 'very poor'. The next stage is to translate each performance parameter into a single ordinal scale. The number of divisions in this scale can be chosen to suit the amount of information and time available. A five-point scale could be used to obtain an approximate result, while an eleven-point scale would provide better resolution, if accurate values for the performance parameters can be obtained. Such an extended scale would require more thought, and therefore more time. The final stage of the method involves allocating weighting values to each of the evaluation criteria, possibly by using the objectives tree method discussed above. The product of the ordinal score and the appropriate weighting value is then entered into each of the matrix

Criterion	Conventional design (benchmark)	New concept
Range of light positions	0	−1
Ease of positioning	0	0
Attractive appearance	0	+1
Ease of manufacture	0	+1
Total score	**0**	**+1**

Figure 8.5 A concept evaluation matrix, using as an example the desk lamp concept derived earlier.

The concept evaluation matrix is used to highlight the strengths and weaknesses of alternative concepts, and to generate numerical scores indicating their overall merit.

cells. Because the weighting values represent the relative importance of the various criteria, the cell entries can be summed for each column to give an overall 'utility value' for each of the different design concepts.

Whichever method is used, the scores achieved by the competing concepts can be used to identify the 'best' design. Alternatively the information in the matrix can be used to identify weak points in each of the selected concepts, and so find ways of improving them. The amended design concepts can then be re-evaluated in the same way as before.

Figure 8.5 shows a concept evaluation matrix using the first of the two approaches described above. The matrix provides an evaluation of the desk lamp concept described in the previous section, using the more conventional three-joint, balance sprung design as a benchmark. Although it does well elsewhere, the new concept is worse than the conventional design with respect to the range of light positions. This criterion is arguably the most important of those listed in the matrix, but this priority is not reflected in the final score, which indicates that the new concept is an improvement on the conventional design. The second type of concept evaluation matrix would reflect more accurately the relative priority accorded to the various criteria.

Taguchi method

The aim of the Taguchi Method (Taguchi, 1986) is to reduce variability in the performance of the product being designed. The approach focuses on a single aspect of the product's performance, the 'customer-related parameter', for which there is a target value or a range of acceptable values. Variability in the value of this parameter is inevitable, due in part to the finite tolerances specified for products and their components, and in part to expected variations within the operating environment.

For success the Taguchi Method depends on close co-operation and interaction between design and production functions to identify noise factors (both design- and production-related) which influence the variability of the customer-related parameter. Optimal values for these parameters are found by conducting experiments in which each factor is varied in turn. A significant part of the method lies in a system for designing this experimental programme so that the required information is obtained from the smallest possible number of experiments.

Value engineering

Much design work is concerned with varying and adapting existing products, rather than the creation of new designs. Indeed, as much as 90–95 per cent of design work may be adaptive or variant in nature. No matter how minor the change that is made to an existing product design, the goal of variant design activity is to seek out an improvement over the previous solution and, according to Cross (1989), all such modifications can be classified into one of two types:

- Those aimed at increasing the value of the product to the purchaser.
- Those aimed at reducing the cost of the product to the manufacturer.

Consequently design activity, whether it be variant, adaptive or original, is primarily concerned with adding value. The value engineering method focuses a designer's attention on increasing the value of a product and/or reducing the cost. This differentiates it from value analysis, which is simply concerned with cost reduction. Again according to Cross (1989), the basic procedure to be observed is as follows:

1. List the separate components of the product, and identify the function served by each component.
2. Determine the values of the identified functions. These must be the values as perceived by customers.
3. Determine the costs of the components. These must be after they are fully finished and assembled.
4. Search for ways of reducing cost without reducing value, or of adding value without adding cost.
5. Evaluate alternatives and select improvements.

Value engineering activities are normally labour intensive and therefore time consuming. They are often performed by a multidisciplined team.

Design for manufacture

Designers cannot produce manufacturable designs without some knowledge of the manufacturing processes through which their designs will be realised; methods of ensuring that the design teams acquire this knowledge were discussed in Chapter 4. Failure to take the possibilities, constraints and costs associated with these production processes into account when designing a product will result in unnecessarily high costs and lead times. Conversely, costs and lead times can be minimised by making a concerted effort to consider production issues as an integral part of the design activity. This idea has been accorded the acronym DFM, design for manufacture.

Designing a product involves making decisions regarding the number of component parts, their materials and geometry, and how they fit together. Although the function of the product is usually the primary issue in these interrelated decisions, careful consideration of production issues can lead to lower costs for both component production and assembly, in addition to shorter cycle-times and better product quality.

Component parts are made using a variety of production processes. Different processes are suited to different materials, impose different constraints on part geometry, and imply different relationships between material and geometry on the one hand and production quality, time and cost on the other. For example, the geometry of a machined component is likely to be restricted to a combination of simple shapes such as cuboids and cylinders. No such restriction applies if the component is injection moulded, but the shape of the part should be chosen with attention being paid to how the material will flow into the mould, and how the formed part will be removed from it. Considering factors such as these when designing a product is known as design for piece-part producibility, or DFP.

The process of assembling components can also be carried out using a number of different technologies, and again these different technologies impose different requirements and result in different costs. Considering assembly processes during the design of a product is known as design for assembly, or DFA (Andreasen, 1988), which Pugh (1990) relates to a product complexity factor, defined in terms of the number of parts, the number of types of part and the number of interconnections and interfaces between parts. This analysis of complexity originated in the field of electronic system design, where it is claimed that lower complexity results in better reliability, lower cost and higher quality. This, and other research (Boothroyd and Dewhurst, 1987), has shown that assembly costs can be reduced by eliminating non-essential components, and by minimising the number of types of parts (i.e. part standardisation). Complexity analysis can be applied at quite an early stage in the design process, with rough comparisons of assembly costs for alternative design solutions being made at the sketch or preliminary layout stage. The beneficial effect of this activity is likely to be greater than that arising from extensive cost-reduction exercises at later stages in design.

Decision matrices

The relative attractiveness of competing design solutions often depends on factors which are outside the designer's control. In this type of situation the uncertainty surrounding these externally imposed factors can make the process of choosing between solutions difficult. Decision matrices (Asimow (1962); Simon, (1975)) can aid this process by clarifying the information involved in the decision, and by externalising the associated thought processes.

The first step in constructing a decision matrix is to select a number of possible scenarios regarding the externally imposed factors. Consider an example where the price of a particular material, which varies between 70 per cent and 130 per cent of a datum value, has a critical effect on the relative costs of two alternative design solutions. After choosing some distinct points on this scale of prices (perhaps 80 per cent, 100 per cent and 120 per cent) the designer can construct a decision matrix in which each row corresponds to one of the design solutions and each column corresponds to one of the chosen scenarios. For each design, the cost corresponding to each scenario can then be calculated and inserted into the appropriate matrix cell, as shown in Figure 8.6.

The information contained in a decision matrix can be used in different ways. One approach, known as the 'minimax criterion', is to

Price of material (relative to datum)	80%	100%	120%
Unit cost of design A	$8,500	$9,000	$9,500
Unit cost of design B	$7,000	$8,500	$10,000

Figure 8.6   An example decision matrix, showing the effects of possible material price variation.

The decision matrix is used to illustrate the consequences of alternative design choices in the event of different circumstances. In this case the price of a particular material impacts significantly on the relative costs of the two alternative designs.

choose the solution which has the lowest maximum cost. This is a risk-averse strategy, which would be adopted if the consequences of a poor outcome were unduly severe. Another approach is to assign probabilities to the scenarios and calculate the weighted sums of the cost values for the different solutions. This probability-based approach is more evenly balanced between opportunities and risks, and should be used if the losses from poor outcomes can be offset by gains from good outcomes.

Computer-based tools

This section describes some computer-based tools that may be of use in design. The number-crunching power of computers has been utilised in engineering for many years, and many computationally intensive engineering analysis tasks can now be carried out using commercially available software packages. Computer-based analysis tools are of great value to designers, as they save time and reduce the chances of errors in analysing and evaluating designs. However, the aim of this section is not to cover the vast range of available analysis tools, but to focus on some computer-based tools whose application in design is in the area of solution synthesis and selection. The process of design synthesis is an elusive subject to describe, and is therefore a challenging area for the developers of design software. However, progress is being made on these types of tools, which will be the design tools of the future.

Optimisation methods

The design process may be viewed as a search for the best possible solution to a given problem. A measure of quality, therefore, must be established in order to identify which design is 'best'. Minimum cost is the most common measure, but in some design problems other criteria such as minimum weight or minimum energy consumption may be considered more important. Whichever criterion is used, the value of the criterion is an outcome of many decisions made during the process of designing the product. The more design decisions there are, the greater the number of possible design options, each one of which is potentially the 'best'.

If the number of possible options is small it may be possible, by using analytical models, to calculate for each one the corresponding value of the criterion to be optimised. However, this course of action is clearly not feasible if the number of options is large, which is often the case. Numerical optimisation methods provide an alternative means of identifying optimal designs in this situation. Due to the complex and repetitive nature of the calculations involved, these methods are invariably used in the form of computer-executed algorithms. These methodically manipulate the design variables in a computer model of the system, as shown in Figure 8.7. Some engineering analysis software packages, for tasks such as chemical process simulation and structural steelwork design, have built-in optimisation facilities. If a design problem can be modelled using one of these packages, then this is undoubtedly the easiest way to gain access to the benefits that optimisation methods can offer. However, many general purpose optimisation packages are also available, either as public domain software or from technical and scientific software houses.

Which optimisation package to use depends on the nature of the design problem. The principal choice is between packages that use 'gradient-based' or 'hill-climbing' methods (Gill *et al.*, 1981; Scales, 1985) and those that use semi-random methods such as 'genetic algorithms' (Holland, 1975). If the design problem can be formulated in terms of locating the best set of values for a number of real-valued design variables, then a gradient-based algorithm will prove effective in many cases. However, these algorithms cannot be used for problems which involve choices between discrete alternatives, or where values must be selected for non-continuous variables. This limitation does not apply to genetic

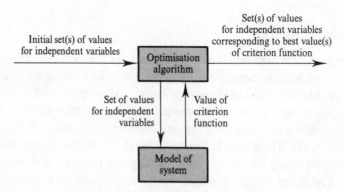

Figure 8.7 Optimisation in design.
Optimisation algorithms work in conjunction with a model of the design, so that the relative success of each proposed alternative design can be evaluated. The optimisation process searches methodically through the possible values of the independent variables until the 'best' design is identified.

algorithms, making them an appropriate choice for many optimisation problems in design.

Knowledge-based systems

Knowledge-based systems, or expert systems, are computer-based systems which can be used in both synthesis and analysis applications. They are a branch of artificial intelligence and have found commercial success in analytical and diagnostic-type applications. Essentially, they can be used to assist in the design process by offering advice, or making decisions, on questions which would normally be addressed by a number of experts.

The reason that knowledge-based systems have found success in analytical and diagnostic applications is that it is much easier to make explicit the knowledge that is used to evaluate or solve a problem than the knowledge that is required for synthesising solutions. As such, knowledge-based systems are particularly useful for addressing well-defined or well-structured problems. Those systems that have found commercial success have tended to be in applications where the computer can be used to take intelligent decisions and then justify its own line of reasoning in a manner which is intelligible to the user. Two particularly interesting examples of knowledge-based systems have recently been released in the UK. The first system, produced by Lloyd's Register of Shipping, consists of a computerised database comprising all of the detailed engineering knowledge previously encapsulated in the many volumes which represented the organisation's rules and regulations for ship design. The system comprising this substantial body of knowledge has an operator-friendly graphical user interface which designers can use to interrogate the validity of their designs without having to make reference to the original weighty texts. The second example, produced by the Health and Safety Executive (HSE) in the UK, is a system which encapsulates all known information and data relating to the prevention of dust explosion and the sizing of explosion vents. As such, the system is a valuable tool which designers working in the powder handling industries can use to analyse and subsequently refine their process plant designs.

Research is continuing on systems that will also allow ill-defined problems to be solved. Indeed, one of the major goals in the artificial intelligence community is to develop intelligent systems that are capable of interacting with the external world by using a broad spectrum of problem-solving and learning methods. For example, recent work has focused on the development of a software environment that can assist in designing, amongst other things, high-rise office buildings (Newell and Steier, 1993).

The most difficult aspect of creating a knowledge-based system lies in eliciting the knowledge and expertise that resides only in the brains of human experts. If this can be achieved, by a process of skilful interrogation, it represents the transfer of that knowledge from human to computer memory. There are two obvious benefits derived from this. First, the expertise is available at all times, and could be made available anywhere in the world, and second, the expertise is not lost when any one individual retires. In some fields this preservation of knowledge is important as the skill base in many industries is being reduced through retirement and redundancy.

Decision support systems

Designing is a problem-solving process in which available information is used to make decisions regarding the preferred form of a product. In many design problems the nature of the required information and the structure of the necessary decisions are not entirely understood before the problem-solving process begins. These design problems fall into the category of 'unstructured' or 'unprogrammable' problems (Simon, 1960, 1969). They are differentiated from 'structured' or 'programmable' problems, which use a predefined sequence of operations to derive values for a predefined set of output data from a predefined set of input data. Two examples of structured problems are payroll production and finite element analysis, for which computers have been used for many years. However, as the power and flexibility of computers have increased, it has become increasingly possible to develop interactive computer-based systems for solving semistructured and unstructured problems. These sophisticated problem-solving tools are called decision support systems, or DSS (Bennett, 1983).

Rather than providing a guided tour through a preprogrammed problem-solving process, a DSS allows users to make up the process as they go along. The DSS helps them to explore the structure of the problem in hand by providing a range of representations (such as graphs, charts and tables) through which they can view the relevant information. The DSS also provides operations which support the three major problem-solving activities of problem definition, elaboration of alternatives, and evaluation and selection. Having arrived at a particular representation of the problem in hand, a user can choose from a number of different operations, each of which will result in a new representation. The solution to the problem is arrived at through a repeated process of gathering and analysing information. This approach to problem-solving parallels that outlined at the beginning of Chapter 7.

Much of the work in the area of DSS has been conducted from the standpoint of management decision-making and management information systems, and little work has been done in applying DSS to design problems, despite the clear parallels between design theory and the principles of DSS. Although there is scope for using DSS in design, at this stage it is still an experimental design tool.

Summary

A designer's tool kit does not only contain the tools that are necessary for the process of designing. As well as managing information within the design process, designers have to manage other resources, and to interact with the people and agencies that directly influence their working environment. For this reason designers also need tools for project management, project planning, document storage and retrieval, and so forth. On a day-to-day basis, however, it is through the use of design tools that product ideas are first generated, and so it is these tools which define the role of the designer. It is therefore the ability of designers to use these tools effectively that determines the quality of the end product, and correspondingly the quality of the world in which we live.

Conclusion

In bringing together the various elements of this text, the authors have tried to satisfy several objectives. First, to provide a useful introductory guide to the field of engineering design, covering the context within which design activity typically takes place, a brief history of design practice and theory, and some of the key activities that a designer may become involved with. Second, to provide a comprehensive reference list to key studies, and other sources of material in the field, which can provide further information and insights into specific topics. Finally, to promote the idea that design itself offers the opportunity for a stimulating and rewarding career which can yield immense personal satisfaction, as well as bringing satisfaction to others. Although the material contained in the text is not an exhaustive treatise on the subject of design, it does serve as a primer and can therefore be employed to introduce many of the other more extensive and complex issues that are involved. It should be of benefit both to academics and industrialists, and in particular prove to be suitable material for use on engineering courses at degree level or equivalent.

As was stated in the Introduction, the topics contained in this text were selected on the basis that each represents a different perspective of design, the intention being to inform the reader as to the nature of design by providing a rounded view. In the process two issues have been raised repeatedly, and so perhaps deserve some final comments.

Consider first the issue of the continuing differences between design models and design practice. Many of the prescriptive models of design reflect a problem-focused approach to generating successful solutions. In contrast, much design practice follows a product-focused approach, as echoed in the descriptive models. Hence there is a clear difference between what is thought to actually take place, and what is considered (by some) to be best practice. Neither of these views can be described as wrong, since both approaches are employed in industry and are producing good new products every day. It is not the intention in this discussion to

146

show favour to one or other of these approaches, but it should be pointed out that a different approach may need to be employed for different types of problem-solving activity. Moreover, the best approach may depend on the nature of the resources available to perform the activity, and on the emphasis placed on innovation by the company. Clearly it is sensible to commence all design activity by designing the design process for the particular problem, taking account of the time and resources available. In this context the observation of Asimow is also relevant. He demonstrated that in any particular case the approach adopted to solving a problem is likely to influence the outcome. Therefore, while it is necessary to begin a design by deciding on the most appropriate approach, this can only be identified after information about the problem has been gathered.

Although a dichotomy continues to exist between prescriptive design models and design practice, evidence suggests that industry has gradually been taking many of the ideas of the theorists on board and applying them where it sees fit. It would be untrue to claim that designers in industry have wholeheartedly adopted all the ideas, but where the benefits of applying a particular aspect of an approach have been firmly established, changes in working practice have been implemented.

A second issue raised repeatedly in the text is the move in industry toward multidisciplined teams, and the changing role of the designer in this context. The subject of concurrent processes (i.e. simultaneous engineering) has become prominent in recent years. To date, however, many of the models in the literature have tended to reflect the idea that design is a discrete activity, which can be performed in isolation from other activities. This view is to some extent justifiable, and has been brought about by the tendency to decompose whole processes, in both design and manufacturing, into a series of sequentially linked tasks. The emergence in manufacturing of this notion (perhaps two hundred years ago) initiated a movement which until fairly recently found much favour. It is interesting to note, however, that the situation has virtually turned full circle, and many engineering organisations are now returning to a philosophy based on integrated team effort. Whereas in the past it was the craftsman who conceived and created products, today it is frequently, and rightly, the responsibility of a multidisciplined team. Under this new regime, designers represent only a part of a team. It is therefore important that the theorists do not ignore both the interactions that must occur between designers and other members, and the fact that designers are tending to take on extra responsibilities within the team.

The driving forces behind the shift to this new philosophy are clear. As was stated earlier, one of the prime factors is that some problems have become so complex and intricate that a single person could never be expected to generate a successful solution. Another significant motivating factor is that, in order to be successful, businesses today must be able to respond swiftly to rapidly changing market conditions and requirements. New technical disciplines such as electronics, and other scientific developments such as those in materials science, are significantly influencing many aspects of society. These rapid shifts have dictated that the team approach is in many cases essential. Boeing, for example, use a multidisciplined multilocation team to design their extraordinarily complicated product, the modern jet airliner. However, the use of a team as opposed to a single resource does create other difficulties. Issues such as the management and control of the team effort, cultivation of the right team dynamic and the need for ever-improving communication between team members, all have to be addressed. (Boeing make extensive use of the latest facilities available in information technology in order to keep all team members abreast of developments during the very complex design process.) Because of the nature of this new team approach some of the models of design, while still valid within certain limits, do not encompass the holistic perspective a designer needs to adopt today. Any future contributions to the field must take account of this perspective.

As the procedures in engineering design change, so the influence and responsibilities of the design function are shifted. Although design is still a sub-function of a much larger process, the distinctions between design and the other disciplines are becoming blurred. The scope of the designer's role is expanding, and this demands that designers of today must be multitalented, possess good interpersonal skills and be able to operate effectively in teams. The changing climate also requires designers to broaden their skill base, so as to encompass such activities as project management and planning. The aims of the manufacturing business have not changed, but the demands being placed on designers have. To deliver the required results, designers of tomorrow must embrace the challenge of their widening role.

Bibliography

Akao, Y. (1990), *Quality Function Deployment: Integrating Customer Requirements into Product Design*, Productivity Press, Cambridge, MA.

Alexander, C. (1971), 'The State of the Art in Design Methods', *Design and Manufacturing Group Newsletter*, 5, No. 3, p. 3.

Andreasen, M. M. (1988), *Design for Assembly*, IFS Publications Ltd, Bedford.

Andreasen, M. M. (1991), 'Design Methodology', *Journal of Engineering Design*, 2, No. 4, p. 321.

Andreasen, M. M. and Hein, L. (1987), *Integrated Product Development*, IFS Publications Ltd, Bedford.

Archer, L.B. (1965), *Systematic Method for Designers*, Design Council, London.

Asimow, M. (1962), *Introduction to Design*, Prentice Hall, Englewood Cliffs, NJ.

Bahrami, A. and Dagli, C. H. (1993), 'Models of Design Processes', in H.R. Parsaei and W.G. Sullivan (eds), *Concurrent Engineering: Contemporary Issues and Modern Design Tools*, Chapman & Hall, London.

Bendell, T. (ed.), (1989), *Taguchi Methods: Proceedings of the 1988 European Conference*, Elsevier Science Publishers, Barking.

Bennett, J. L. (1983), *Building Decision Support Systems*, Addison-Wesley, Reading, MA.

Boothroyd, G. and Dewhurst, P. (1987), *Product Design for Assembly*, Boothroyd & Dewhurst Inc., Wakefield, RI.

Braun, A. (1980), 'Interdependence between Social and Technical Innovations', in B.-A. Verdin (ed.), *Current Innovation – Policy, Management and Research Options*, Almqvist & Wiksell International, Stockholm.

Braverman, H. (1974), *Labour and Monopoly Capital*, Monthly Review Press, New York.

Brereton, M., Cupal, T. and Leifer, L. (1993), 'Reconstructing reality in engineering exercises: an experience in design, manufacture and use', *Proceedings of the International Conference on Engineering Design*, The Hague, Netherlands, 17–19 August 1993.

British Standards Institution (1987), *BS5750: Quality Systems*, British Standards Institution, London.

British Standards Institution (1989), *Guide to Managing Product Design, BS7000: 1989*, British Standards Institution, London.

British Standards Institution (1991), *Guide to the Preparation of Specifications, BS7373: 1991*, British Standards Institution, London.

Broadbent, G. and Ward, A. (eds) (1969), *Design Methods in Architecture*, Lund Humphries, London.

Cooper, R.G. (1984), 'The Performance Impact of Product Innovation Strategies', *European Journal of Marketing*, **18**, No. 5, pp. 5–54.

Cornfield, K. G. (1979), *Product Design*, National Economic Development Office, London.

Court, A. W., Culley, S. J. and McMahon, C. A. (1993), *A Survey of Information Access and Storage amongst Engineering Designers*, Internal Report No: 008/1993, School of Mechanical Engineering, the University of Bath.

Coyne, R. D., Rosenman, M. A., Radford, A. D., Balachandran, M and Gero, J. S. (1990), *Knowledge-Based Design Systems*, Addison-Wesley, Reading, MA.

Cross, N. (1989). *Engineering Design Methods*, John Wiley & Sons, Chichester.

Cross, N. (1993), 'Science and Design Methodology: A Review', *Research in Engineering Design*, **5**, p.63.

Cross, N. and Roozenberg, N. (1992), 'Modelling the Design Process in Engineering and in Architecture', *Journal of Engineering Design*, **3**, No. 4, pp. 325–337.

Cross, N., Dorst, K. and Roozenburg, N. (eds) (1992), *Research in Design Thinking*, Delft University Press, Delft.

Darke, J. (1978), 'The Primary Generator and the Design Process', in W. H. Ittelson *et al.* (eds), *EDRA9 Proceedings*, University of Arizona, Tucson.

Dixon, J. R. (1966), *Design Engineering: Inventiveness, Analysis and Decision Making*, McGraw-Hill, New York.

Dixon, J. R. (1987), *Artificial Intelligence in Engineering Design and Manufacture*, **1**, No. 3, p. 145.

Eekels, J. (1983), 'Design Processes Seen as Decision Chains: Their Intuitive and Discursive Aspects', in *Proceedings of the International Conference on Engineering Design, ICED '83, Copenhagen*, Vol. 3, Heurista, Zurich.

Erkens, A. (1928), *Beitrage zur Konstruktion Serziehung*, Zeitschrift VDI, 72, 1928, pp.19–21.

Evans, B. J., Powell, J. and Talbot, R. (eds) (1982), *Changing Design*, John Wiley, Chichester.

Feilden, G. B. R. (chairman, the Feilden Commission), (1963), *Engineering Design*, HMSO, London.

Fowler, T. C. (1990), *Value Analysis in Design*, Van Nostrand Reinhold, New York.

French, M. J. (1985), *Conceptual Design for Engineers*, The Design Council, London.

Fricke, G. (1993), 'Empirical investigation of successful approaches when dealing with differently precised design problems', *Proceedings of the*

International Conference on Engineering Design, The Hague, Netherlands, 17–19 August 1993.

Gero, J. (ed.) (1991), *Artificial Intelligence in Design '91*, Butterworth-Heinemann, Oxford.

Gill, P. E., Murray, W. and Wright, M. (1981), *Practical Optimization*, Academic Press, London.

Glegg, G. L. (1969), *The Design of Design*, Cambridge University Press, New York.

Gobeli, D. H. and Brown, D. J. (1987), 'Analysing Product Innovations', *Research Management*, July–August 1987, pp. 25–31.

Hajek, V. G. (1977), *Management of Engineering Projects*, McGraw-Hill, New York.

Hales, C. (1987), *Analysis of the Engineering Design Process in an Industrial Context*, Grants-Hill, Cambridge.

Hales, C. (1993), *Managing Engineering Design*, Longman Scientific & Technical, Harlow.

Hall, A. D. (1962), *A Methodology for Systems Engineering*, Van Nostrand, Princeton, NJ.

Hartley, J. and Mortimer, J. (1991), *Simultaneous Engineering: The Management Guide to Successful Implementation*, Department of Trade and Industry/Industrial Newsletters Ltd, Dunstable.

Hillier, B., Musgrove, J. and O'Sullivan, P. (1972), 'Knowledge and Design', in W. J. Mitchell (ed.), *Environmental Design: Research and Practice*, University of California Press, Los Angeles.

HMSO (1994), *Competitiveness: Helping Business to Win*, HMSO, London.

Holland, J. H. (1975), *Adaptation in Natural and Artificial Systems*, University of Michigan Press, Ann Arbor.

Hollins, W. and Pugh, S. (1990), *Successful Product Design*, Butterworth, London.

Holt, K. (1983), *Product Innovation Management*, Butterworth, London.

Hubka, V. (1982), *Principles of Engineering Design*, Butterworth, London.

Hubka, V., Andreasen, M. M. and Eder, W. E. (1988), *Practical Studies in Systematic Design*, Butterworth, London.

Jones, J. C. (1970), *Design Methods*, John Wiley, Chichester.

Jones, J. C. and Thornley, D. G. (eds), (1963), *Proceedings of the Conference on Design Methods*, Pergamon Press, Oxford.

Krick, E. (1978), *An Introduction to Engineering*, Wiley, New York.

Kuhn, T. S. (1970), *The Structure of Scientific Revolutions*, Chicago University Press, Chicago, IL.

Lawson, B. (1990), *How Designers Think – The Design Process Demystified*, Butterworth Architecture, Oxford.

Leech, D. T. and Turner, B. T. (1990), *Project Management for Profit*, Ellis Horwood, Chichester.

Levin, P. H. (1984), 'Decision-making in Urban Design', in N. Cross (ed.), *Developments in Design Methodology*, John Wiley, Chichester.

Littler, C. R. (1985), *The Experience of Work*, Gower, Aldershot.

Littler, C. R. (1986), *The Development of the Labour Process in Capitalist Societies*, Gower, Aldershot.

Love, S. (1980), *Planning and Creating Successful Engineered Designs: Managing the Design Process*, Advanced Professional Development, Los Angeles, CA.

Luckman, J. (1984), 'An Approach to the Management of Design', in N. Cross (ed.), *Developments in Design Methodology*, John Wiley, Chichester.

McGrath, M. E., Anthony, M. T. and Shapiro, A. R. (1992), *Product Development: Success through Product and Cycle-Time Excellence*, Butterworth-Heinemann, Boston, MA.

Maffin, D. (1996), 'Engineering Design and Product Development in a Company Context', PhD Thesis, University of Newcastle, Newcastle upon Tyne.

March, L. J. (1976), *The Architecture of Form*, Cambridge University Press, New York.

Meyer, M. H. and Roberts, E. B. (1986), 'New Product Strategy in Small Technology-Based Firms: A Pilot Study', *Management Science*, 32, No. 1.

Nevins, J. L. and Whitney, D. E. (1989), *Concurrent Design of Products and Processes*, McGraw-Hill, New York.

Newell, A. (1969), 'Heuristic Programming: Ill-structured Problems', in J. Aronofsky (ed.), *Progress in Operations Research*, John Wiley, New York.

Newell, A. and Steier, D. (1993), 'Intelligent Control of External Software Systems', *Artificial Intelligence in Engineering Design and Manufacture*, 8, No. 1, pp. 3–21.

Oakland, J. S. (1989), *Total Quality Management*, Butterworth-Heinemann, Oxford.

Oakley, M. (1984), *Managing Product Design*, Weidenfield & Nicholson, London.

Offshore Engineer (1991), 'Wall Failure Suspected for Sleipner Failure', *Offshore Engineer*, October 1991, pp. 21–24.

Open University (1993), *T363 – Computer Aided Design, Units 1/2 – Introduction to Computer Aided Design*, Open University, Milton Keynes.

PA Consulting Group (1989), *Manufacturing Into the Late 1990s*, Department of Trade and Industry, HMSO, London.

Pahl, G. and Beitz, W. (1984), *Engineering Design*, The Design Council, London.

Papalambros, P. and Wilde, D. (1988), *Principles of Optimal Design: Modelling and Computation*, Cambridge University Press, New York.

Popper, K. R. (1963), *Conjectures and Refutations*, Routledge & Kegan Paul, London.

Pugh, S. (1986a), 'Design Activity Models: Worldwide Emergence and Convergence', *Design Studies*, 7, pp. 167–173.

Pugh, S. (1986b), *Curriculum for Design: Specification Phase*, SEED (Shared Experience in Engineering Design), Hatfield.

Pugh, S. (1990), *Total Design*, Addison-Wesley, Wokingham.

Reuleaux, F. and Moll, C. (1854), *Konstruktionslehre Fur den Maschinenbau*, Vieweg, Braunschweig.

Rittel, H. and Webber, M. (1973), 'Dilemmas in a General Theory of Planning', *Policy Sciences*, 4, p.155.

Ross, P. J. (1988), *Taguchi Techniques for Quality Engineering*, McGraw-Hill, New York.

Scales, L. E. (1985), *Non-linear Optimization*, Macmillan, London.

Simon, H. A. (1960), *The New Science of Management Decisions*, Harper & Row, New York.

Simon, H. A. (1969), *The Sciences of the Artificial*, Massachusetts Institute of Technology Press, Cambridge, MA.

Simon, H. A. (1975), *A Student's Introduction to Engineering Design*, Pergamon Press, New York.

Simon, H. A. (1984), 'The Structure of Ill-structured Problems', in: N. Cross (ed.), *Developments in Design Methodology*, John Wiley, Chichester.

Smith, D. G. and Rhodes, R. G. (1992), 'Specification Formulation – An Approach that Works', *Journal of Engineering Design*, 13, No. 4.

Smith, P. G. and Reinertson, D. G. (1991), *Developing Products in Half the Time*, Van Nostrand Reinhold, New York.

Suh, N. P. (1990), *Principles of Design*, Oxford University Press, Oxford.

Taguchi, G. (1986), *Introduction to Quality Engineering*, Asian Productivity Organization, Tokyo.

Takeuchi, H. and Nonaka, I. (1986), 'The New New Product Development Game', *Harvard Business Review*, 64, Jan–Feb, pp. 137–146.

Tovey, M. (1986), 'Thinking Styles and Modelling Systems', *Design Studies*, 7, No. 1, p. 20.

Twiss, B. (1986), *Managing Technological Innovation*, Pitman Publishing, London.

Ullman, D. G. (1992), *The Mechanical Design Process*, McGraw-Hill, New York.

Verein Deutscher Ingenieure (VDI) (1987), *VDI-2221: Systematic Approach to the Design of Technical Systems*, Beuth Verlag, Berlin.

Wallace, K. M. and Hales, C. (1987), 'Detailed Analysis of an Engineering Design Project', *Proceedings of ICED 87: International Conference on Engineering Design*, Boston, MA., 17–20 August 1987.

Walsh, V., Roy, R., Bruce, M. and Potter, S. (1992), *Winning by Design: Technology, Product Design and International Competitiveness*, Blackwell, Oxford.

Wheelwright, S. C. and Clark, K. M. (1992), *Revolutionizing Product Development: Quantum Leaps in Speed, Efficiency and Quality*, The Free Press, New York.

Index

Bold page numbers denote figures.

Index